A Short History of the Lincolnshire Red Shorthorn Cattle

by Geo. E. Collins

with an introduction by Jackson Chambers

Self Reliance Books

Get more historic titles on animal and stock breeding, gardening and old fashioned skills by visiting us at:

http://selfreliancebooks.blogspot.com/

Introduction

I am pleased to present another title in the "Cattle" series.

The work is in the Public Domain and is re-printed here in accordance with Federal Laws.

As with all reprinted books of this age that are intended to perfectly reproduce the original edition, considerable pains and effort had to be undertaken to correct fading and sometimes outright damage to existing proofs of this title. At times, this task is quite monumental, requiring an almost total "rebuilding" of some pages from digital proofs of multiple copies. Despite this, imperfections still sometimes exist in the final proof and may detract from the visual appearance of the text.

I hope you enjoy reading this book as much as I enjoyed making it available to readers again.

Jackson Chambers

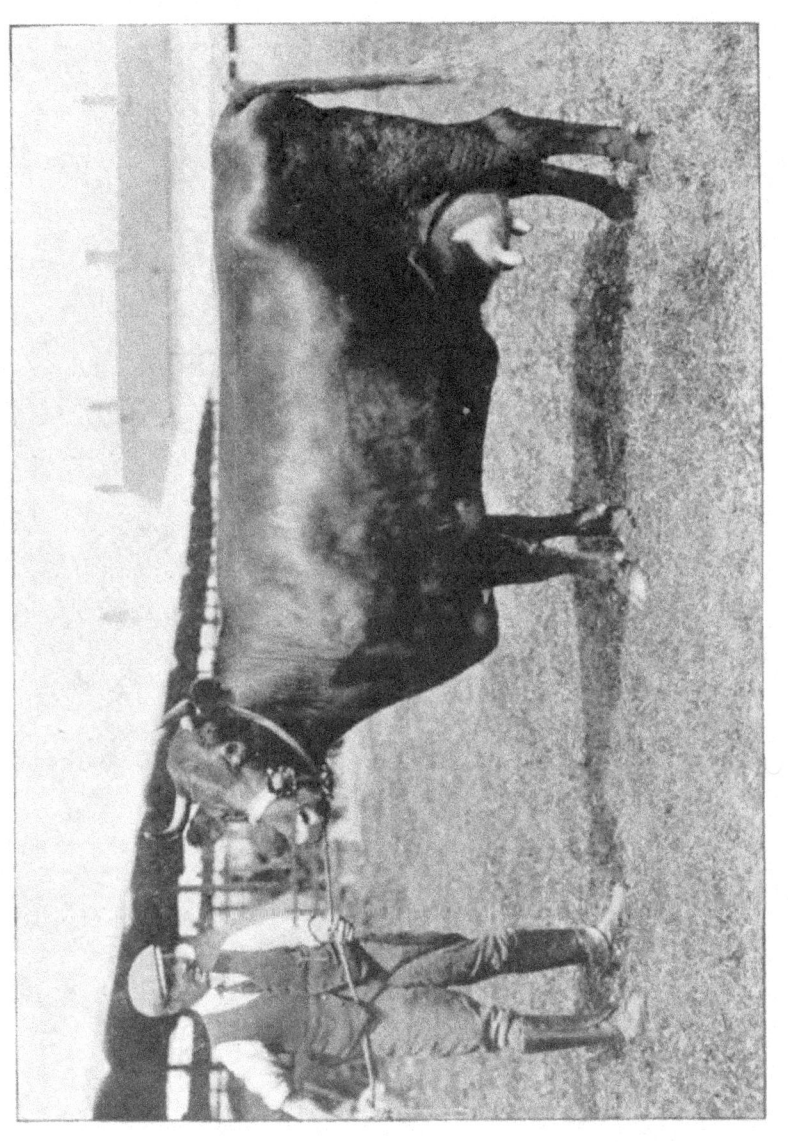

BURTON QUALITY 3RD,

Bred by and the property of Mr. JOHN EVENS, Burton, Lincoln,
Champion Lincoln Red Cow in 1907 and 1908, Reserve Champion in 1909.

THIS SHORT HISTORY

OF THE

LINCOLNSHIRE RED SHORTHORNS

IS DEDICATED TO MY FRIEND

GEORGE MARRIS,

OF

KIRMINGTON HOUSE, BROCKLESBY,

with whom I learnt farming and first became

acquainted with this typical general purpose,

tenant-farmers' breed of cattle

THE AUTHOR

THE EARLY HISTORY

OF

LINCOLNSHIRE SHORTHORNS

AND THE FOUNDATION OF THE

LINCOLN REDS.

THE EARLY HISTORY OF LINCOLNSHIRE SHORT-HORNS AND THE FOUNDATION OF THE LINCOLN REDS.

The tenant farmer of to-day has to exercise every possible economy if his balance sheet of accounts is to be a satisfactory one at the end of each year, and so he cannot be too careful in the selection of the cattle with which to stock his farm ; for he cannot generally afford to ride a hobby-horse, and so he must turn to the class of animal which he believes will prove a real rent-payer. As the only true test of the value of cattle can be obtained from the block and milk-pail, the class of animal that produces the greatest amount of valuable meat for the butcher, and the class whose females give the largest amount of milk of the required standard must surely be the best breed for the tenant-farmer. It used to be said :— " Booth for the butcher, and Bates for the pail." But the average tenant-farmer requires an animal that will combine both these qualities. And this the admirers of the Lincoln-shire Red Shorthorns claim for their breed.

The tenant-farmer, too, must have a class of cattle that is hardy and thrifty, is possessed of sound constitutions, requiring no cosseting, but one rather that is able to with-stand much inclement weather and rough treatment, and a class that quickly comes to maturity and produces an early profit. The history of the Lincoln Reds shows that no breed is more suited to the farmers' requirements. Wintered in fold-yards with little or no shelter ; fed on barley straw and a few turnips ; exposed to the coldest of winds and the wettest of weather, the weakest have been weeded out with most marvellous certainty. About the middle of April they are turned out to get their own living, facing the biting east winds from the North Sea, and if there is any delicacy in cow or calf, it is soon discovered. During the hot months of summer there is often little drink but such as is provided by stagnant ponds. Such is the test that has been going on in the case of the Lincoln Reds for a hundred years and more. And the result ? Distinctly a case of the survival of the fittest ; a race of cattle that not only do not lose their condition under circumstances that would have had the most disastrous results as regards most breeds, but thrive and grow on and improve, laying on a wealth of lean flesh, and providing a bountiful supply of rich milk for the nourishing of their off-spring. This is a state of things that tenant-farmers all over England should take note of, this superiority to adverse conditions, this ability to grow on and develop where other cattle would pine and shrink. The cattle markets of the East and Midland Counties, where steers at two-and-a-half years old may be looked upon to yield seven or eight cwt. of the best meat, speak most eloquently of the beef-producing qualities of the Lincoln Reds ; and the butchers will tell you that they cut up a far greater proportion of lean flesh, with very much less offal, than any other breed they know. By judicious selection and under proper treatment the great milking capabilities of the breed can easily be developed, as has been proved by the truly wonderful successes of the Burton herd, near Lincoln, belonging to Mr. John Evens ; for in the premier milking trials, both in England and Ireland, he has demonstrated the superiority of the Lincoln Reds over all other breeds. His example has been followed by Mr. F. Scorer, of Bracebridge and Nettleham, near Lincoln, whose herd is fast acquiring a reputation in dairy trials, and other breeders are developing the great milking qualities of the breed. And so the reputation of the Lincolnshire Red Shorthorns as a dual-purpose cattle is now fully established, for no other breed can boast of such beef-producing steers and at the same time such milk-producing females.

The impression the cattle made at the Royal Show in 1901, when they were first granted seperate breed classes by the Royal Society, was a distinctly favourable one ; and the general opinion of agriculturists and stock-breeders from all parts of the kingdom was that they were bred to a well-defined type, that they showed great wealth and evenness of flesh, and that their milking qualities were undeniable. But they have made great strides since then, as was evidenced at the Royal Show, held at Lincoln in 1907, and at each Royal and Lincolnshire Show since, where the cattle have invariably made a display of which their breeders had every reason to be proud. By going further afield to Shows and taking mental notes of what they saw, and by greater care of their cattle and more judicious breeding at home, the result of the lessons taught by the Show ring, the breeders of the Lincoln Reds have now a class of cattle still possessing all their old good qualities but neater in appearance, shorter on the leg, and with better backs and more true Shorthorn character than when they first made their bow to the public at Cardiff, in 1901. But one point the breeders of the Lincoln Reds must be careful to guard against. It is most desirable that greater neatness and more Shorthorn character should be aimed at, but it should not be at the expense of the size and substance for which the breed has always been so famous. An occasional infusion of C.H.B. blood is often desirable, but care should be taken that such an infusion does not lead to weaker constitutions, impaired milking capabilities, and a loss of size and substance. The dairy countries of Europe have long been good customers ; South America is taking larger numbers every year ; and South Africa promises to be one of its best markets, for no other breed has fulfilled the requirements of the country so nearly as do the Lincoln Reds. Registered herds are springing up all over England outside the county boundaries ; and on dairy farms from all parts comes an increasing demand both for bulls and heifers. Built on Shorthorn lines— with great length and scale, and typical heads—the chief characteristics, then, of the Lincoln Reds are their early maturity, hardiness and thriftiness, great wealth of lean flesh, and splendid milking qualities.

The Preface to the first volume of the Lincolnshire Red Shorthorn Association's Herd Register states that the Association was formed (in 1895) to promote the interests of this variety of the great Shorthorn race by publishing a register, and securing uniformity of type and colour. It states that the original cattle of Lincolnshire in their unimproved state were distinguished by their enormous size, but slow powers of fattening, and that about 100 years before the date of the Preface their improvement was commenced by the introduction of the new type of Shorthorn which then arose in Durham and Northumberland. Reference is also made to the bulls sent into Lincolnshire from Charles Colling's great sale in 1810. It states that one of the most potent factors in bringing Lincolnshire Shorthorns to their present type was the herd formed by Mr. Thomas Turnell, at Reasby, near Wragby, towards the close of the last century. Quoting from Arthur Young's remarks upon Lincolnshire in his report to the Board of Agriculture, the preface says :—" Mr. Turnell has a breed of cattle which are not surpassed by any in the county for points highly valuable, or their disposition at any age to fatten rapidly. His bull covers at a guinea, and has many cows sent to him. The breed originally came from the neighbourhood of Darlington." He describes the cattle as of medium size, which he preferred to larger ones. It is unfortunate that more minute records do not exist of the methods of breeding pursued by Mr. Turnell, but most of the best herds in the county at the present time acknowledge the influence of the "Turnell Reds." Mr. Turnell impressed his cattle with the deep cherry red colour now so much the fashion, and while slightly reducing their size from the original type, gave them greater powers or rapid fattening, and of developing the primest joints of meat. Mr. Coulam, of Withern, Mr. Baumber, of Somersby, Mr. Oliver, of Eresby, whose herds are all dispersed, and Mr. Cartwright, of Tathwell, whose herd was sold in 1844, maintained the excellence of the breed in the earlier part of the present century.

Turnell Reds were introduced into the neighbourhood of Bourne, in the extreme south of the county, by Lord Willoughby de Eresby, and Mr. Redmile, of Dyke, who bought cattle from Mr. Oliver, of Eresby.

It would be invidious, says the preface, to select any existing herds for special notice, but no account of Lincoln Red Shorthorns can be complete which does not mention the share the Messrs. Chatterton have had for two generations in maintaining the character of these cattle up to the present day especially by two most successful out-crosses they took in sending a famous cow bred by Mr. Coulam, of Withern, to Lord Exeter's Cambridge Duke 5th (C.H.B. 30,644), and in later times by sending one of their best cows to Mr. Deane Willis' Windsor Benedict (C.H.B. 40,933), the result in each case being most impressive, the bulls from these two unions having done great service to the breed. Messrs. Burtt, of Welbourne, have also maintained one of the oldest existing herds of these cattle, which under the name of the " Old Welbourne Reds," have had much influence in the south and centre of the county.

The preface describes the Lincolnshire type of shorthorn as being distinguished by its great length of frame, good constitution, great hardiness, capacity for milk, and great weight of carcase, 8 to 10 cwts. being usual weight for grass-fed bullocks, and up to 24 cwts. for stall-fed show cattle.

Records of the leading herds, though not entered in Coates' Herd Book, have been kept in some cases for nearly 100 years, and the breed has gradually conformed to one type and colour, as shown at the yearly sales of bulls held at the Lincoln April Fair. Freedom from the restraint of a paper pedigree increases the usefulness of a breed during its period of development, because it enables breeding animals to be selected for merit alone, and not because they spring from one particular line of blood.

Dairying, says the preface, is not a prominent feature in Lincolnshire agriculture, and the practice of allowing cows to suckle their own calves is not conducive to the development of milking capacity; but whenever proper attention is paid to the encouragement of milk, Lincolnshire Red Shorthorns show that they can yield an abundant supply. The wonderful success of the breed at dairy shows, milking trials and butter tests proves the truth of these remarks.

Indeed, since this preface was written, the dairy trials at the Royal Show have demonstrated that as regards weight of milk and general average of points the Lincoln Reds are far ahead of all other breeds. From the date of the formation of the Association it was decided to conduct the annual sale of bulls at the Lincoln April Fair under the auspices of the Association.

Referring to the Marquis of Exeter's bull, Cambridge Duke 5th (C.H.B. 30,644), it might be mentioned that the Burghley Park herd was one of the oldest in the kingdom, and had been bred by the Marquises of Exeter for many years. Entries were made in the first volume of Coates' Herd Book, and bulls were used from Warlaby, and Mr. Unthank's herd. In 1864 the famous bull, Fourth Duke of Thorndale was purchased from Mr. Hales, for 410gs., and a blend of the old Gwynne blood with the Burghley Innocence lines produced the renowned bull Telemachus, who was the first bull to win over £1,000 in prize-money at shows. Between 1868 and 1880, this herd won upwards of £4,000 at the leading shows.

The subscribers to the Memorandum of Association, which is dated May 17th, 1895, were :—

Mr. James Hornsby, Laxton Park, Stamford.
Mr. Edmund Turnor, Panton Hall, Wragby.
Mr. C. W. Tindall, agent to the Brocklesby estate.
Mr. J. H. Dean, Greatford, near Stamford.
Mr. T. B. Freshney, South Somercotes, Louth.
Mr. Robert Chatterton, Stenigot.
Mr. Walter Martin, Wainfleet.

The following were the officers elected to serve on the first Committee of Management :—President, James Hornsby,

Esq.; Vice-president, Edmund Turnor, Esq.; Treasurer, A. S. Leslie-Melville, Esq.; Members of Committee, W. R. Caudwell, E. H. Cartwright, Robert Chatterton, J. H. Dean, John Evens, T. B. Freshney, Walter Martin, G. E. Sandars, W. Scorer, C. W. Tindall, W. W. Wells, W. Wingfield; Secretary, Mr. Stephen Upton; Auditor, Mr. W. K. Broughton.

Mr. James Sinclair in his graphic and interesting " History of Shorthorn Cattle " says that all our domesticated cattle are clearly descended from two pre-historic types :—Bos urus (called alternatively Bos primigenius) and Bos longifrons. These two types and their modified descendants differed mainly in size,—the urus being of gigantic dimensions, and the longifrons comparatively small, and known in Britain as the Celtic shorthorn ox.   The urus is not believed ever to have been domesticated in Britain, though it was known here in a wild state in prehistoric times.   But when the Romans landed in the year 55 B.C. they found the longifrons type domesticated there ; and Julius Cæsar left it on record that " the number of its cattle was great."   But by this time the urus was quite extinct in Britain, although there is no doubt that a larger type of cattle than the longifrons, probably descended from the urus, existed in Europe about that time, they in turn being descended, there is little doubt, from the domesticated type of urus which existed in Asia, whence it must have spread westward.   It is believed that the gradual increase in the size of the British cattle was due to the importation of this modified urus type by the inhabitants of Jutland, Holstein, and Friesland when they invaded the country between 449 and 660 A.D., an invasion which was very different to that of the Romans, who came as fighting men, with fresh conquests to make ; whereas our forefathers, the English, were distinctly colonists, and brought their wives and children, cattle and *lares et penates* generally.   To these people must date our improved system of farming, and the larger type of cattle from which all our more valuable breeds are descended.

Gervaise Markham, in his book, " A way to get wealth," published in 1695, wrote, in speaking of the choice of " a fair bull," that the best of our English cattle are bred in Yorkshire, Derbyshire, Lancashire, Staffordshire, Lincolnshire, Gloucestershire and Somersetshire.   Those bred in Yorkshire, Derbyshire, Lancashire and Staffordshire, he said, were generally all black in colour, while those in Lincolnshire for the most part ' pyde,' with more white than the other colours ; " their horns little and crooked, of bodies exceeding tall, long and large, lean and thin-thighed, strong hoved, not apt to sorbate, and are indeed fittest to labour and draught."

George Culley in " Observations on Live Stock," published in 1785, says there is little doubt that " the Short-horned or Dutch kind " was imported from the Continent, first because they are in many places still called the Dutch breed, and secondly because very few of them were to be found except along the Eastern coast of the Island, facing that part of the Continent where the kind of cattle was still bred. " In Lincolnshire," says Culley, " which is the furthest south that one meets with any number of this breed of cattle, they are, in general more subject to lyer or black flesh than those bred further north, and in that rich part of Yorkshire called Holderness they are much the same as those of which we have been speaking." In a later edition of his book, published in 1792, Culley remarks, that in a journey through Lincolnshire in 1784, he was happy to find that many sensible breeders had improved their breed of Shorthorn cattle very much since his tour in that country ten years previously, by using good bulls and heifers brought from the counties of Durham and York, on both sides of the Tees, where the best were confessedly bred. The cattle imported from the Continent were no doubt of the type represented in the pictures of Paul Potter, Rubens, Cuyp and Teniers of the unimproved Shorthorn type. In Mr. John Thornton's paper on " Ancient Shorthorns," he states that the late Mr. William Torr, of Aylesby Manor, Lincolnshire travelling in Holland, in 1838, visited Utrecht Fair and saw a large number of animals fully resembling ordinary Shorthorns ; they were rare milkers, had tolerable formation, a good skin, mellow handle, and nice waxy horns and with every variety of colour. It must be remembered that the Dutch did not create the Shorthorn breed, but they helped to improve it by introducing greater milking qualities. The importation of Dutch cattle for crossing with the Shorthorn had wholly ceased before the time of the Collings, about the beginning of the last quarter of the 18th century. The birth of the modern Shorthorn may certainly be said to date from the Collings, Robert who lived at Barmton and Charles who lived at Ketton, in the county of Durham. They founded their system of breeding on that learnt from Mr. Robert Bakewell, at Dishley, and the real improvement in the Lincolnshire Shorthorn dates from the introduction of their blood into the county, where, at that time, there were some good cattle, more or less under the influence of the Dutch cross.

The great hit that was responsible for the fortunes of the Ketton herd was the purchase, mainly through the influence of Mrs. Colling, of the cow Favourite (afterwards called Lady Maynard), and her calf, Young Strawberry, by Dalton Duke (188) ; and thus was laid the main source of the Shorthorn breed of the future for the sum of *thirty-five pounds for cow and calf.* Mr. Maynard, of Eryholme, at first refused all offers made him by Mr. Colling, and even Mrs. Colling's bid was considered for a time ; "a killing bid" although it was supposed to be. Mr. Charles Colling's bulls used to be let out by the year at 50gs. and 100gs. each, and among his hirers were Mr. Ostler of Aylesby and Audleby. At the sale at Ketton, on October 11th, 1810, Mr. Grant of Wyham purchased the cow Laura, by Favourite (252), whose dam was the before-mentioned Lady Maynard (Favourite), Laura being out of another good cow, Lady, and her price being 210gs. ; and Mr. Grant also gave 106gs. for Lady's bull calf, Lucilla, by Comet (155), and 200gs. for the bull Major (397). Mr. Grant used to cross his Turnell cattle with Colling-bred bulls, and this herd is undoubtedly one of the foundations of the Lincoln Reds of the present day. Mr. P. Skipworth of Aylesby, was also a purchaser at the above sale, paying 140gs. for Young Favourite (254), by Comet (155) out of Countess. As Favourite and Comet had so much to do with the building up of the Lincolnshire Shorthorn a description of them will be of interest. Mr. J. C. Maynard describes Favourite (252) as a "beautiful animal ; so beautifully made and possessing such a wonderfully noble style of look and walk, very masculine too and active. He had bronze horns, large and as thick as my arm nearly, and one of the finest heads I ever saw." Of Comet (155), by Favourite (252) from Phœnix, Mr. George Coates said " I never saw his equal," and Mr. Charles Colling himself declared him to be the best bull he ever bred or saw. " He was a beautiful light roan, dark neck, with a fine masculine head, broad and deep breast, shoulders laid back, crops and loins good, hind quarters long, straight and well packed, thighs thick, twist full and well let down, with nice straight hocks and hind legs. He had fair-sized horns, ears large and hairy, and a grandeur of style and carriage that was indescribable."

More cattle went into Lincolnshire from Robert Colling's sale at Barmton, on September 29th and 30th, 1878, among the number being Lily, by North Star, from a Favourite cow, who went to Mr. P. Skipworth, Aylesby, at 66gs. ; Vesper, by Wellington, from another cow, by Favourite (252), who was bought by Mr. J. White, Coates, at 111gs. ; Princess, by Lancaster, from Golden Pippin, also knocked down to Mr. P. Skipworth, at 156gs., who likewise bought Violet, by North Star from a cow by Midas, at 48gs. ; Cowslip, by Wellington out of a Favourite dam, who was sold to Mr. Leighton, North Willingham, at 54gs. ; and the bull Major, who became the property of Mr. W. Brooks, Laceby, at 185gs., he being by Wellington, out of a Phenomenon cow.

At Mr. Christopher Mason's sale at Chilton, Durham, on August 31st and September 1st, 1829, the bull Childers, by Satellite out of a cow by Cato, was bought by Mr. Dudding to go into the Panton herd, at 225gs. ; and numerous purchases were made at the sale of Mr. Thomas Bates' herd at Kirklevington, on May 9th, 1850, among the number being Wild Eyes 19th by Second Duke of Oxford (9046) out of Wild Eyes 10th, who was sold to Mr. Cartwright, Haugham, Louth, at 60gs. ; Wild Eyes 22nd, by Second Cleveland Lad (3408) out of Wild Eyes 8th, who was knocked down to Mr. H. Champion, Ranby House, for 100gs., who also bought Duchess 62nd, by Second Duke of Oxford, from Duchess 56th, at 120gs., while Mr. Cartwright also secured Wild Eyes 27th, by Second Cleveland Lad out of Wild Eyes 17th, at 43gs.

A Lincolnshire herd of note in the first half of the 19th century was that belonging to Mr. W. Smith, West Rasen, who entered animals in the first volume of the Herd Book, and no fewer than 28 bulls bred by him are entered in the first three volumes. A sale of stock bred by him was held in September, 1852, when 73 head realized £2,814. Mazurka, a yearling by Harbinger, went first into Mr. H. Ambler's herd, near Halifax, and then out to the United States, where she founded a tribe there—the Mazurkas—which long enjoyed a favour among American breeders hardly inferior to that won by the best Kirklevington cattle.

Mr. W. Torr, of Aylesby Manor, was one of the best farmers and one of the greatest breeders of Shorthorns who ever lived in England. He was a born judge, knew good cattle when he saw them, and being entirely free from bigotry and prejudices, founded a herd equal to any in the kingdom. He was a great admirer of Booth cattle, and in the forties considered that Mr. John Booth's herd at Killerby, was unquestionably the best small herd in England. The best herd of over 100 head he said he ever saw, belonged to Mr. C. Robson, of Cadeby, from whom he bought the founders of his famous M. and G. families. His own maxim in breeding was " cull the worst and never sell the best females, no matter how tempting the price." When the herd came to be dispersed at Aylesby, on September 2nd, 1895, 84 herd of cattle were disposed of for £42,919 16s., and an average of £510 19s. The sale is a landmark in Shorthorn history and was in its way unique, as every animal offered was bred on the farm by the late owner.

Another gentleman who took a keen interest in Shorthorn breeding was Mr. J. Banks Stanhope, whose herd was dispersed at Revesby Abbey, in 1862. His cattle came of a great dairy stock, and the herd contained many of Mr. J. Beasley's old J. family, as well as Knightly Walnuts and Barringtons. At the sale 48 head averaged £41 9s., Bracelet going to Mr. G. T. Heneage, at 150gs.

The Dudding family formely lived at Panton, near Wragby, where a herd of cattle and flock of Lincoln sheep were kept upwards of a century and a half ago. A picture of " The Panton Heifer," bred by Mr. John Dudding, painted and engraved by H. Calvert, is still preserved in the family. In 1860 a sale of Shorthorns took place, when an average of £51 4s. 6d. was obtained for 97 herd. Mr Henry Dudding's annual sale of Shorthorns and Lincoln sheep, at Riby Grove, near Grimsby, is now one of the principal events in the live stock year. Twelve sales have been held, at which 661 head of Shorthorns have been sold to average £53 2s. 3d. and 558 shearling rams, £68 2s. 7d., the total proceeds of the twelve sales (including 373 shearling ewes, which averaged £8 5s. 10d.) being £76,219 4s. 9d. Mr. Dudding has recently been offered and has refused the sum of £38,000 for his flock, herd, and tenant-right.

It will now be seen on what a sure foundation the Lincoln Red breed of cattle was built up. But for some reason or other the bulk of the breeders refrained from registering their herds in Coates' Herd Book, and gradually conforming to one type and colour by using none but red bulls from herds of renowned constitution within the limits of the county ; and only dipping into C.H.B. blood by introducing red bulls for an out-cross ; they eventually developed a distinct breed of Red Shorthorns which, though, lacking somewhat of the neatness of the C.H.B. animal, was claimed to possess more hardiness and thriftiness, an ability to fatten quicker and come to earlier maturity, and with greater milking qualities than the parent breed. And so in 1895 came the formation of an Association, with a Herd Register, at the instigation of Mr C. W. Tindall, of Wainfleet (then agent to the Earl of Yarborough, at Brocklesby Park). Since then greater attention has been paid to appearance, while careful to lose none of the old qualities ; and there are now 339 members of the Association, and the bulls registered in the Herd Book number 7,286. Very few C.H.B. herds now remain in the county, those belonging to Mr. H. Dudding, Riby Grove ; Mrs. Webb & Sons, Melton Ross ; Messrs. S. E. Dean & Sons, Dowsby Hall ; and Mr. J. E. Casswell, Laughton, being among the most prominent ; but 98 per cent. of the cattle bred in the county are Lincolnshire Reds, registered or unregistered.

B

THE CATTLE IN THE SHOW RING.

THE CATTLE IN THE SHOW RING.

1896. In *1895* it was decided by the Breed Association to offer prizes for animals either entered or eligible for entry in the Lincolnshire Red Shorthorn Herd Book, at the Lincolnshire Agricultural Society's Show at Gainsborough, in *1896*. There were three classes—for bulls under two years old, for cows and calves, and for heifers under two years old; and there were two prizes in each class of the value of £10 and £5 respectively, and of these prizes £25 was given by Mr. J. D. Sandars, Gainsborough, and £15 by the Association. In the bull class the first prize went to Mr. Pereira Brown, Glentworth Hall, Lincoln, for Rear-Admiral (920), by Admiral (3), bred by Mr. J. Reeson, Keddington, Louth. Admiral, who was bred by Mr. G. Griffin, West Ashby, and was the property of Mr. E. H. Cartwright, Keddington, was by Commodore (81), bred by Mr. W. Chatterton, Hallington, Louth. The Earl of Yarborough, Brocklesby Park, took the second prize with the home-bred Brocklesby Albert (587), also by Admiral (3), from a cow by Orangeman (190). The reserve card went to Lord Yarborough for First-Lieutenant (374), by Stentor (255).

Messrs. T. & J. B. Freshney, South Somercotes, Louth, took premier honours for cow and calf with Ruby 2nd, by Saltfleet Eclipse (227), from Ruby 1st, by Saltfleet South Ormsby (229). Mr. William Scorer, Sudbrooke, came next with Sudbrooke Beauty, by a bull bred by Mr. Stafford, of Marnham, and from a cow bred by Mr. King, of Ormsby, she being by a Hallington bull. Mr. Walter Martin, Wainfleet, and Lord Yarborough, stood reserve and commended respectively, the latter showing that famous cow Nonpareil, by Hyllus (149), and she had at her heels a bonny bull calf, by Volunteer (C.H.B. 67,364) afterwards known as Nonsuch (1292), when the property of Mr. W. Chatterton, Hallington. At the head of the heifers stood Dowsby Milker 2nd, the property of Messrs. S. E. Dean & Sons, Dowsby Hall, Folkingham, she being by Cambridge Duke 30th (C.H.B. 60,441), from Dowsby Milker, by Edenham King (112); next to her being Messrs. T. & J. B. Freshney's Ruby 3rd, by Saltfleet Eclipse (227) out of Ruby 1st.

1897   At the County Show, at Sleaford in *1897*, two prizes
were given in each of five classes, £20 being offered from
the Sleaford local fund and £35 by the Association.   There
were two classes for bulls, the first being for two-year-olds and
upwards, and at the top of the class the judges placed a bull
belonging to Mr. R. Davy, Worlaby, by Archer (16), who was
by Hallington (135), out of a cow by Cawkwell (67).   Archer
was bred by Mr. G. E. Sandars, Fillingham, and Hallington,
by Mr. W. Chatterton, Hallington, while Cawkwell was bred by
Mr. C. B. Robson, Cawkwell House, Horncastle.   Both Hallington and Cawkwell belonged to Mr. G. E. Sandars.   The
second place was occupied by Red King (480), belonging to
Messrs. R. & R. Chatterton, Stenigot, he being by Red
Prince (211), from Stenigot Red Rose 2nd, by Kinsman 21st
(C.H.B. 37,206), his great grand-dam being by Hyllus (149).
Baron Kelby, by County Member (83), the property of Mr. T.
Barrand, Kelby, Grantham, stood reserve, and commended
cards were handed to two bulls who afterwards became sires
of great note, and left their mark on the Lincoln Red breed.
One of these was Bigby (319), the property of Mr. E. H. Cart-
wright, Keddington, Louth, and bred by Mr. G. Walker,
Bigby, Brigg.   He was by Orangeman (190), a Keddington
bred bull, and his dam was the celebrated Nonpareil, by Hyllus
(149).   The other was Chancellor (332), bred by Mr. Walter
Martin, Wainfleet, and the property of Mr. G. Langham, Hough
Manor, Grantham.

The bulls under one year old were led by Messrs. R. &
R. Chatterton's Captain of the Guard (667), by Corporal (688).
He was bred by Mr. W. Hyde, Market Stainton, his dam being
by Candidate (56), by Butterman 3rd (C.H.B. 44,487).   Cor-
poral was by Commodore (81), and he was bred by Messrs.
Chatterton at Stenigot.   The afterwards famous sire Nonsuch
(1292), stood next.   He was the property of Mr. W. Chatterton,
Hallington, and was bred by the Earl of Yarborough, being by
Volunteer (C.H.B. 63,501), from Nonpareil, she being by Hyllus
(149), from a cow bought at the Stenigot dispersal sale by a
bull belonging to the late Mr. W. Chatterton, and was bred by
Mr. John Chapman, South Field, Louth.   Thurlby Cambridge
Duke (1001), by Cambridge Duke 30th (C.H.B. 60,441),
belonging to Mr. T. A. Bettinson, Thurlby Manor, Bourne,
and Dowsby Lincoln Waterloo (368), by Dowsby Waterloo
Duke (102), the property of Colonel M. W. Willson, C.B.,
Rauceby Hall, Sleaford, were respectively reserved and commended.   Messrs. R. & R. Chatterton's Stenigot Bloom, by

Candidate (56), from Bloom, by Hyllus (149), bred at Stenigot, obtained premier honours in the class for cows with calves at foot ; the second ribbon being awarded to Mr. J. Langham for Keddington Satellite, by Eclipse (111), from Keddington Lady. This Eclipse, which, owing to his registered number being 111, frequently and erroneously appears in print as Eclipse the Third, was one of the landmarks in Lincoln Red breeding. He was bred and owned by Mr. W. Chatterton, Hallington, his sire being Windsor Benedict (C.H B. 40,933), and his dam Luna, by Hercules (144). Windsor Benedict was bred by Mr. T. Willis, Carperby Manor, Bedale, Yorks., coming of Warlaby (Mr. Booth's) blood, and he belonged to Mr. E. H. Cartwright, Keddington ; and Hercules, who was bred by the late Mr. W. Chatterton, Hallington, was by Cambridge Duke 5th (C.H.B. 30,644), from Alcama, whose dam came from the famous herd belonging to Mr. Coulam, Withern. Mr. E. H. Cartwright's Keddington Favourite, by Admiral (3), from Keddington Butterfly, by Worlaby Lad (289), her dam Keddington Chrysalis, by Windsor Benedict (C.H.B. 40,933), was reserved. Heifers not over two-and-a-half years old were led by Messrs. R. & R. Chatterton's Stenigot Buttercup 5th, by Red Prince (211), from Stenigot Buttercup, by Candidate (56), her dam Buttercup, by Hallus (149). Red Prince was by Commodore (81), from an Eclipse (111) cow, and was bred by Mr. W. Chatterton. Mr. Langham's Keddington Satellite again occupied the second place, and Messrs. Dean & Sons' Dowsby Milker 2nd, by Cambridge Duke 30th (C.H.B. 60,441), from Dowsby Milker, by Edenham King (112), was reserved. At the top of the heifers under 18 months old stood Vanity Gwynne, the property of Messrs. R. & R. Chatterton, she being by Coastguard (343), out of Stenigot Vanity by Comet (79), grand dam Vanity 3rd, by Prince Gwynne 3rd (C.H.B. 43,799), great-grand-dam Vanity, by Hyllus (149). Comet was bred by Mr. W. Chatterton, and was by Eclipse (111), out of a cow by Wellington Duke (C.H.B. 47,243). Red Daisy from the same herd occupied the second place, she being by Commander (80), from Steingot Daisy 2nd by Comet (79), her grand-dam being Stenigot Daisy by Hyllus (149), and her great-grand-dam Daisy by Hercules (144). Commander was by Commodore (81), out of a candidate (56) cow. The reserve number fell to a heifer by Volunteer (C.H.B. 63,501), the property of Mr. C. W. Tindall, Wainfleet, and bred by the Earl of Yarborough.

1898.   The following year, *1898*, the Lincolnshire Agricultural Society held their annual exhibition at Lincoln, and there were again five classes. The Stenigot herd was once more very much to the fore in the prize list, beginning with a victory in the class for bulls two years old and upward with Volunteer (1715), a home-bred animal. He was by Red Prince (211), out of Stenigot Red Rose 3rd, who in turn wasy Crœsus by (85), out of Red Rose, she being by Prince Gwynne 3rd (C.H.B. 43,799) out of a cow by Hyllus (149) ; he was afterwards sold to go to Argentina. Red Prince was bred at Hallington, being by Commodore (81), from an Eclipse (111) dam ; and Crœsus was a Stenigot bred son of Hyllus (149). The second prize was awarded to Asterby Red (19), bred at Hallington and the property of Mr. T. Bett, Benniworth, and was another of Eclipse's (111) sons ; while the reserve place was occupied by Mr. J. Higgins' Tealby Hero (544), by a bull bred by Mr. Isaac Sharpley, Calcethorpe, and bred by Mr. W. Drakes, Tealby. At the top of the class for bulls one year old stood Beggerman (1457), by King Hal (156), from a Keddington cow, belonging to Mr. G. E. Sandars, Fillingham Manor, Lincoln, his sire being a son of Eclipse (111) and a famous cow called Riches, in the herd of the late Mr. W. Chatterton. The second prize was awarded to another landmark in Lincoln Red breeding, a bull that has proved one of the most successful sires in the history of the breed, and this was Keddington Ruby, (1243). He was shown by Mr. John Searby, Croft, Wainfleet ; but he was bred by Mr. E. H. Cartwright, at Keddington, and he later on went into the possession of Mr. G. E. Sandars, Scampton, Lincoln. His breeding was as good as any in the Herd Book, for he was by Bigby (319), who was by Orangeman (190), from that famous cow Nonpareil by Hyllus (149), and his dam, Keddington Skipworth II. was by Commodore (81), out of Warlaby Skipworth, by Ludford (172), who was a son of Cawkwell (67), and was bred by Mr. G. E. Sandars, Fillingham Manor, and her grand-dam was Haugham Skipworth, by Windsor Benedict (C.H.B. 40,933). The reserve ticket went to Messrs. R. & R. Chatterton for Commander-in-Chief (1497), by Commander (80), out of Stenigot Cowslip, by Comet (79), out of Cowslip, whose dam was by Hercules (144).

The cows with calves at foot were led by Messrs. Chatterton's Stenigot Butercup 5th, by Red Prince (211), out of Stenigot Buttercup, by Candidate (56) ; and next to her stood Grateful, by King Crœsus (155), out of an Eclipse (111) cow, shown by Mr. S. E. Dean, Threekingham, Folkingham. Messrs.

S. E. Dean & Sons stood reserved with Dowsby Withern, by Withern Jumbo (283), while the previous year's second prize winner, Mr. J. Langham's Keddington Satellite had to be content with a commendation. The first three places among heifers under two and a half years of age, were occupied by Stenigot animals, the two prizes going to Vanity Gwynne, who won as a heifer under eighteen months old the year before; and to Countess by Commander (80), out of Stenigot Poverty 2nd, by Admiral (3). Red Daisy, by the same sire as the last, but out of Stenigot Daisy 2nd, by Comet (79), was reserve. The class for heifers not exceeding one year and six months old was headed by Fledborough Gem, by Anderby Hector (298), out of Willoughby Daisy, her sire being by Hector (142) —who was by Eclipse (111), and was bred at Hallington— whose breeders were Messrs. J. N. Robinson & Sons, Anderby Bank, Huttoft. The second and reserve cards went to Messrs. R. & R. Chatterton for Rose of Summer, by Stenigot Knight 2nd (527), out of Stenigot Red Rose 4th, and Stenigot Bloom 3rd, by Wolseley (1436), from Stenigot Bloom.

1899. In *1899* the County Show was held at Louth, the local fund contributing £15, and the Association £45 to the two prizes in each of the five classes. Premier honours in the class for bulls two years old and upwards went to Mr. E. H. Cartwright's famous bull, Bigby (319), now seven years old, but very massive and even in flesh, and good in his lines still. Next to him stood Messrs. T. & J. B. Freshney's Grandad (1561), by Daddy (1507), who came of the late Mr. W. L. Mason's Keddington blood. Baron Ormsby 3rd (26), by Stamp End (247), a Hallington bred bull, was reserved ; his dam was descended from the good strains in the herd of Mr. Oliver, of Fresby, and he was shown by Messrs. S. & J. W. T. Crawley, Hemington, Oundle. But the Stenigot herd came to the front in the class for bulls under two years of age with Sirdar (1676), another bull that has helped to make Lincoln Red history. He was bred by Mr. J. B. Hill, Smethwick Hall, Congleton, Cheshire, and was by Conisholme Boy (347), by Glengarry (C.H.B. 62,653), bred by Mr. R. N. Sutton-Nelthorpe, Scawby Hall, and was out of an Eclipse (111) dam. The second prize fell to Messrs. S. & J. W. T. Crawley for Bumper 2nd (1793), by Baron Ormsby 3rd (26), out of Thurlby Duchess, from Mr. T. A. Bettinson's herd near Bourne. Mr. T. Bett's Saltfleet Actor (1664), by Shooting Star (1674), from a Saltfleet Eclipse (227) dam, was reserve, and Mr. J. Maunsell

Richardson's Healing Baron (1899), by Bigby (319), out of Wainfleet by the Master (C.H.B. 59,456), was commended. The class for cows with calves at foot was headed by Mr. John Evens' Quality, by Collingwood (C.H.B. 57,074), from a cow belonging to Mr. J. Laverack, Carlton, by a bull from the herd of Mr. Stafford, of Marnham. This cow afterwards when mated with Red Rover (C.H.B., 77,618) became the dam of Burton Quality 3rd, for many years a great prize-winner and the Champion Lincoln Red Cow in 1907 and 1908. Another Burton cow came next, Buttercup by Lincoln (C.H.B. 53,111), out of a dam by Badminton (C.H.B. 41,002) ; while the reserve and commended cards went to two of Messrs. R. & R. Chatterton's cows, Stenigot Blush 2nd by County Member (83), out of Stenigot Blush by Crœsus (85), who goes back to Blush Rose by Hyllus (149), and Stenigot Duchess 3rd by Ballywatler (23), from Stenigot Duchess by Hyllus (149). Mr. C. H. Stafford's Fledborough Gem, the winner of the young heifer class the year before, this time topped the heifers under two and a half years old, and she was followed by a heifer by Saltfleet Sappy (502), belonging to Messrs. T. & J. B. Freshney ; and the Stenigot heifers Rose of Summer and Stenigot Choice 3rd were reserve and commended. Messrs. Chatterton also took the first prize for heifers under 18 months old with Stenigot Duchess 4th by Wolseley (1436) from Stenigot Duchess 2nd by Hallington (135), her dam Stenigot Duchess by Hyllus (149) ; the second prize going to a heifer by Brocklesby Albert (578), shown by Mr. J. Walter, Hatton, Wragby. In the reserve place came Weston Pride 6th by Pippin's Pride (C.H.B. 71,157), out of Weston Pride shown by Messrs. W. J. Atkinson, Weston St. Mary, Spalding.

Mr. Evens' Profitable, by Professor (200), from Burton Old Profit by Beauty Bull was second in the open dairy class ; Quality was his cow second at the Derbyshire County Show, where the young bull Royal Burton (1996), by Knight of Chewton (C.H.B. 68,867), secured a second prize ; a similiar prize going to him at the London Dairy Show. The Burton bull Red Rover (213), of Mr. G. E. Sandars' strains, was a first prize-winner at the London Dairy Show ; and as an illustration that this famous milking herd can produce beef as well as milk, Gwynne and Moderate were second and third at the Lincoln Fat Stock Show.

1900. The Lincolnshire Agricultural Society's Show was held at Spalding in *1900*, and again five classes were allotted to the Lincoln Reds, the Spalding Local Fund contributing £15, and the Lincolnshire Red Shorthorn Association £60. In the older bull class Mr. J. Langham's Chancellor (332) at last succeeded in carrying off a first prize, his solitary opponent being Keddington Saturn (803), by Astronomer (21), from Keddington Stamp by Wolarby Lad (289), bred by Mr. E. H. Cartwright, Keddington, and shown by Mr. J. W. Rowland, Fishtoft, Boston. The younger bulls were led by Mr. John Evens' Royal Burton (1996), by Knight of Chewton (C.H.B. 68,867), out of Dairymaid 2nd by Professor (200), her dam by Ramsden (203). Professor was by Fox (122) who was bred by Mr. R. Burtt, Welbourn, and Ramsden, who was bred by Sir John Ramsden, at Coates, near Gainsborough, was by a bull bred by Mr. E. Paddison, of Ingleby.

Royal Burton afterwards took a first at the Derbyshire a second at the Yorkshire, and a first at the London Shows.

Next in order to Royal Burton at Spalding came a bull by St. Serf (C.H.B. 61,786), shown by Mr. T. Diggle, Thorpe House, Ewerby, and the third prize which was offered this year went to Western Nonpareil King (2068), bred by Mr. W. J. Atkinson, Western St. Mary, and exhibited by Mr. W. J. Measures, Dunsby, Bourne. This latter bull was by Pippin's Pride (C.H.B. 71,157), from that wonderful cow, Nonpareil by Hyllus (149). The reserve number went to a bull by Chancellor (332), out of Keddington Satellite, shown by Mr. J. Langham.

First prize for cow and calf fell to Mr. E. H. Cartwright's Keddington Enderby by Worlaby Lad (289), out of Ludford Enderby, by Ludford Windsor Benedict ; and the second to Mr. John Evens for Burton Rose, by Khartoum II. (153) ; while Burton Fashion, by Fox 2nd (748) from the same herd, was reserve. Rose also took a second prize at the Herefordshire Show at Tring ; and Fashion was placed second at the Yorkshire, second at the Derbyshire, and first at the Cheshire Shows.

The pick of the heifers under two years and six months old proved to be Rosebud, by Brocklesby Albert (578) out of Red Rose by Hatton Red (140), shown by Mr. J. Walter, Hatton, Wragby ; while the second and third prizes went respectively to Messrs. R. & R. Chatterton's Duchess 4th, the winner in the young heifer class the year before, and to Mr. W. J. Atkinson's Western Pride 6th, who stood reserved in the same class. Mr. A. M. Wilson, East Keal Manor, Spilsby, was reserve with a heifer by Lord Rosebery (839). Premier honours in the class for heifers under one year and six months old went to Stenigot for Stenigot Daisy 6th, by Wrangler (C.H.B. 71,901) from Stenigot Daisy 2nd, by County Member (83), out of a cow by Hyllus (149). The second place was occupied by a heifer by Chancellor (332), belonging to Mr. J. Langham, and the third by Mr. W. J. Atkinson's Weston Surprise 2nd by Pippin's Pride (C.H.B. 71,157), from Lincoln Rose ; while as reserve stood Mr. C. H. Stafford's Lady Sarah by Calceby Striver (663) out of Croft Ruby.

Mr. Evens' Pansy, who was awarded second prize in the open Dairy Class at the Lincolnshire Show, also won first prize, together with Sissy, by Professor (200), and Primrose, by Burton Jumbo (46) in the Group Class at the London Show, besides winning at the County Show, the bull Royal Burton (1996) was placed first at the Derbyshire and London Dairy Shows and second at the Yorkshire Show. Mopey also took premier honours for Mr. Evens at the Lincoln Christmas Show, giving another instance of how great milk-producing and great beef-producing capabilities may be developed in the same herd.

1901. In the year *1901* the Royal Agricultural Society for the first time granted breed classes at their exhibition at Cardiff, and a very creditable display was made, though by no means a representative one of the Lincoln Reds ; for many of the principal breeders do not exhibit, while for others the distance was a deterrent. But the general opinion round the ring side, however, was that the quality of the cattle was uniformly good, that they were bred to a well-defined type, were particularly even in their flesh, good to handle, and quite the class of animal to fill the consumers' and the butchers' eye. The milking character of the cows was also most favourably commented on, and altogether they made a highly satisfactory impression. There were four classes, the first being for bulls calved in 1897, 98, or 99, and at the top of this the judges placed Messrs. R. & R. Chatterton's Sirdar (1676), who had won in the class for bulls under two years of age at the County Show, at Louth in 1899. Next to him came Mr. J. Searby's Keddington Ruby (1243), while Mr. John Evens' Royal Burton (1996), the first prize young bull at the County Show at Spalding the previous year, was reserve. Bulls calved in 1900 were led by Cropwell Royal, by Lincoln Sailor (1597), out of Cropwell Ladylike by Leadenham Marksman (825), shown by Mr. J. Marriott, Cropwell Butler, Nottingham ; the second prize going to another famous Stenigot bull, Red Chief (2611), by Sirdar (1676), from Stenigot Red Rose 3rd, who was by Crœsus (85) out of Red Rose, by Prince Gwynne 3rd (C.H.B. 43,799), her dam being by Hyllus (149). Chief Secretary (2160), by Lord Chancellor (1606) from Well Rose 2nd, by Lincolnshire Tom (832), belonging to Sir Robert Wilmot, Bart., Binfield Grove, Bracknell, was reserved.

First prize in the class for cows in milk, calved in 1895, 96, 97, or 98, went to Messrs. R. & R. Chatterton for Stenigot Violet 2nd by Stenigot Knight 2nd (527), out of Stenigot Violet, by Comet (79) ; and the second to Mr. J. W. Farrow, Strubby Manor, Alford, for a cow by Weighty Tom 2nd (558), from a Thumper (550) dam ; while Mr. J. Evens stood reserve with Flossy and commended for Nancy. Premier honours for heifers calved in 1899 or 1900 went to Stenigot for Stenigot Daisy 6th, winner in the young heifer class at the County Show at Spalding the year before, with Mr. J. Langham's Brandon Nonpareil, by Chancellor (332), from Wainfleet Keddington, by Bigby (319), second.

At Newark Show, Mr. John Evens was first in the Dairy Class with Peony ; and Dolly, who was fourth in the Milking Trials at the London Show, also took a third prize for inspection.

At the Lincolnshire Agricultural Society's Show at Brigg the same year, Sirdar (1676) was second for Shorthorn bulls, the property of a member of the Society in the county of Lincoln ; and second prizes both in the open class for Shorthorn four-year-old cows and for Shorthorn dairy cows went to Stenigot Violet 2nd, the first prize in the latter class being carried off by Messrs. T. & J. B. Freshney, with Ruby 4th, by Saltfleet Sappy (502), out of a cow by Saltfleet Eclipse (227). Sirdar (1676) also won in the class for Lincoln Red bulls, two years old and upwards, Keddington Ruby (1243) being second, and Royal Burton (1996) reserve, the Royal decisions thus being upheld. The Royal awards were, however, reversed in the class for bulls under two years of age, Messrs. R. & R. Chatterton's Red Chief (2611) being on this occasion placed in front of Mr. J. Marriott's Cropwell Royal ; and another Stenigot bull, Otby Count (3004), by Dunsby Sentinel (1535), bred by Mr. John Abraham, Otby, was reserved. Bull calves were led by Northborough Cromwell 4th (2587), by Baron Ormsby 3rd (26), out of Thurlby Princess, bred by Messrs. S. & J. W. T. Crawley, Hemington, Oundle, and exhibited by Mr. Everett King, Northborough, Market

Deeping.   The second prize was awarded to Mr. W. Hornsby, Burwell Park, Louth, for a calf by Hainton Freeman (1565), out of a cow by Baron Ormsby 1st (24) ; and Mr. J. Marriott's Cropwell Salesman (2484), by Lincoln Sailor (1597), from Cropwell Beeswing by Senator (948) was reserve.

Messrs. R. & R. Chatterton's Stenigot Violet 2nd and Messrs. T. & J. B. Freshney's Ruby 4th carried off the two prizes in the class for cows in milk or in calf, but the latter, which had been placed in front of the Stenigot cow in the Shorthorn dairy class, now had to give place to the other. Three Burton cows carried off the reserve, highly commended, and commended ribbons for Mr. John Evens in Withern Violet by Marshman (180), Burton Fashion and Burton Cherry, by Khartoum (154).   First prize for heifers under two and a half years of age went to Messrs. R. & R. Chatterton's Stenigot Daisy 6th, the Royal winner, but Mr. J. Langham's Brandon Nonpareil had to give second place to Mr. W. J. Atkinson's Western Surprise 2nd, who had been placed third in the young heifer class the previous year.   Messrs. Chatterton stood reserved for Stenigot Buttercup 7th.   Stenigot Pride 5th also scored for them in the class for heifers not exceeding one year six months old, she being by Wolseley (1436) from Stenigot Pride 2nd by Admiral (3), whose dam was by Crœsus (85). Mr. J. Marriott came next with Cropwell Sunshine by Lincoln Sailor (1597) out of Cropwell Snowball by Leadenham Hero 2nd (435), and the third prize was awarded to Mr. W. J. Atkinson's Weston Princess 2nd by Pippin's Pride (C.H.B. 71,157), out of Western Princess ; with Mr. J. Marriott's Cropwell Glitter 2nd reserve.   It will be noticed that this year six classes were provided for Lincoln Reds at the County Show.

At the London Show the first prize in the group class open to all breeds again went to the Lincoln Reds, the winners being Mr. J. Evens' Primrose, a winner the previous year, Car Fox II. by Burton Jumbo (46), and Maud by Professor (200).   Burton Victorious (2319) by Royal Burton (1996) from Bertha, by Red Moss (210) took a first prize for bulls at the same show.

1902.   There were again four classes for the breed at the Royal Show at Carlisle in *1902*, and six classes were also provided by the Lincolnshire Agricultural Society at their annual exhibition, this year held at Boston.   First prize for bulls calved in 1898, 1899, and 1900 went to Messrs. R. & R. Chatterton's Walmsgate Mate 2nd (1722), a bull of great length and substance, by Baron Ormsby 10th (614), bred by

Messrs. Simons, Sutton-on-Sea ; and the second to Mr. J. W. Measures, Dunsby, Bourne, for Weston Nonpareil King (2068) third at the County Show in 1900, and reserve at the Royal the previous year. He had a good middle and a broad level back. First for bulls calved in 1901, was awarded Messrs. R. & R. Chatterton for Peacemaker (3006) by Sirdar (1676) out of Stenigot Gwynne by Comet (79), out of Ackthorpe Gwynne by Prince Gwynne 3rd (C.H.B. 43,797), her dam being Gwynne by Hyllus (149). Next to him stood Cropwell Sovereign (2485), shown by Mr. John Marriott, a bull of good quality with an excellent middle, by Lincoln Sailor (1597), from Cropwell Snowball by Leadenham Hero 2nd (435) ; and the reserve ticket went to Mr. J. Searby's Calceby Marvel (2453), a youngster of great promise by Matchless (1951), from a Hallington (135) cow, bred by Mr. J. Mason, Calceby Manor. Alford. Stenigot Violet 2nd repeated her performance of the year before by winning the first prize for cows in milk calved in 1896, '97, '98, or '99 ; and next to her came Mr. J. Langham's Brandon Nonpareil, a second prize-winner in the heifer class in 1901. The winner was a cow of beautiful quality and excellent dairy character, and was shown by Mr. John Searby, Croft, who purchased her at the Stenigot sale the previous year. Mr. Marriott's cow had a good middle, and the reserve number, Messrs. Chatterton's Stenigot Choice 4th, was a nice type of cow with an excellent udder. Heifers calved in 1900 or 1901 were headed by Cropwell Sunshine, a second prize-winner at the County Show the previous year, beating Messrs. R. & R. Chatterton's Stenigot Buttercup 8th by Sirdar (1676) out of Stenigot Buttercup 6th by Stenigot Knight 2nd (527), while Mr. J. Langham's Brandon Rosebud was reserved.

Walmsgate Mate 2nd (1722) followed up his victory at Carlisle by winning in the older bull class at the County Show, Boston, Weston Nonpareil King (2068) again being second, with Mr. J. W. Rowland's Keddington Saturn (803) third, and Mr. J. Langham's Brandon's Lord Chancellor (2121) in the reserve place. Croft Kindly (2837), big in frame and good in colour, carried off the first prize for Mr. J. Searby in the class for bulls under two years of age ; he was by Scampton Artist (2328) out of a cow by W.B.W. (557), and was bred by Mr. J. Kiend, Hagworthingham, Spilsby. Mr. Everett King's well-grown sappy bull, Northboro' Cromwell IV. (2587), by Baron Ormsby 3rd (26) from Thurlby Princess, was second ; Kirkby Virtusoo (2947), by Virtuoso (C.H.B. 69,763) out of Kirkby Lady Ormsby, by Baron Ormsby 2nd (25), third ; and Mr. J. Searby's Calceby Marvel (2453), reserve.

Another Royal winner scored again for the Messrs. Chatterton in Peacemaker (3006), who carried off the premier honours in the class for bulls under one year old ; and next in order came Mr. J. Marriott's Cropwell Sovereign (2485) by Lincoln Sailor (1597) out of Cropwell Snowball, by Leadenham Hero 2nd (435) and Nero (2991) shown by Mr. W. Chatterton, Hallington, a promising young bull by Red Monarch (C.H.B. 77,605) out of a cow by Hyllus (149). Mr. W. Chatterton's Numbo (3507) by Nono (2274) out of an Eclipse (111) dam, was reserve.

Brandon Satellite, belonging to Mr. J. Langham, was considered the best of the cows in milk or in calf, she was by Chancellor (332) from Satellite, and was followed by a Weighty Tom 2nd (558) cow belonging to Mr. J. W. Farrow, Strubby Manor, Alford ; and Mr. E. H. Cartwright's Miss Calceby 3rd by Bigby (319) from Miss Calceby by Calceby (51). Mr. J. Evens' Burton Rose by Khartoum II. (153), a third prize-winner at Newark, was reserved. The Royal winner, Cropwell Sunshine, won a first prize for Mr. J. Marriott in the class for two and a half year old heifers ; but Messrs. R. & R. Chatterton's Stenigot Buttercup 8th, who stood next to her at Carlisle, had to go down a place, the second prize being awarded to Mr. E. H. Cartwright's Keddington Pearl by Bigby (319), from Keddington Sapphire by Saltfleet Sappy (502). Mr. J. Langham's Brandon Rose Bud was reserved. At the top of the heifers under a year old the judges put a daughter of Salt-fleet Actor (1664) and a cow by Benniworth Kelstern (28), belonging to Mr. T. Bett, Benniworth Walk, Donington-on-Bain ; and awarded the second prize to Mr. J. Marriott's Cropwell Laura, by Lincoln Sailor (1597), out of Cropwell Ladylike by Leadenham Marksman (825) ; and the third to Mr. W. J. Atkinson's Stenigot Weston Rose by Sirdar (1676), out of Stenigot Red Rose 2nd, bred by Messrs. R. & R. Chatterton. Mr. J. Marriott's Cropwell's Duchess was reserve. In the open Shorthorn classes at the same show, Walmsgate Mate 2nd (1722) and Peacemaker (3006) were each reserved ; Mr. J. W. Farrow's cow was third in her class, and Brandon Satellite reserve ; while Mr. J. Evens' Burton Margaret and Quality and Mr. J. Searby's Stenigot Violet 2nd were all highly commended. Stenigot Buttercup 8th and Burton Ruby 3rd were respectively reserve and highly commended, and Stenigot Daisy 6th was also commended. A Silver Salver for the best milch cow of any age exhibited in any of the Shorthorn or Lincoln Red classes, was awarded to Mr. J. Evens' Burton Margaret.

At Newark Show, Mr. Everett King's Northborough Cromwell 4th (2587) was third, and Mr. W. J. Atkinson's Stenigot Daisy 6th, and Mr. J. Evens' Burton Rose second and third in their respective classes. Mr. J. Langham's heifer, Brandon Satellite won in her class, and Brandon Rosebud was third in another. Mr. Evens also took a third in the Dairy Class at Newark with White Knee II. and a third for inspection at Tring with Margaret, who was the winner in the Milking Trials there, with 76lbs. milk in 24 hours, the highest weight ever given at Tring, the premier Milk Test in the country. This cow was also second in the Butter Test. Young Cherry by Butter Boy (1477) took a second prize back to Burton from the heifer class at the London Show.

1903. At the Royal Show at Park Royal, London, in *1903*, the Lincoln Red classes were increased to six, and six classes were also provided for the breed at the County Exhibition at Lincoln. Red Chief (2611) continued his victorious showyard career for Messrs. Chatterton by winning in the class for three of four-year-old bulls, at Park Royal; the second prize going to Mr. Everett King's Northborough Cromwell 4th (2587), with Mr. John Marriott's Birthday (1782) by Nonsuch (1292) out of a cow by Eclipse (111), bred by Mr. W. Chatterton, Hallington, reserve. The winner showed great wealth of flesh and an excellent middle, in which latter respect the second prize bull was very much like him. The two-year-old bulls were led by Mr. J. Marriott's Cropwell Sovereign (2485), good in colour and even in flesh, who went up one place from the last year at Carlisle; and Messrs. R. & R. Chatterton's Peacemaker (3006) who had beaten him there, had to be content with second honours. The first prize for bulls calved in 1902 went to Mr. Marriott for Cropwell Red Earl (2851), by Lincoln Sailor (1597), out of Cropwell Beeswing by Senator (948). He was a bull of much promise, and was followed by another good young bull in Weston Monarch 2nd (3144) by Red Monarch (C.H.B. 77,605), out of Western Charm by Pippin's Pride (C.H.B. 71,157), shown by Mr. W. J. Atkinson. Messrs. S. E. Dean & Sons, Dowsby Hall, Bourne, were reserved for Dowsby Czar 6th (2875) by Dowsby Czar (C.H.B. 76,574), out of a cow by Cambridge Duke 30th (C.H.B. 60,441).

Mr. John Searby's Stenigot Violet 2nd repeated her performance of the year before by winning in the class for cows and heifers, in milk, calved previous to or in 1900. She was of nice scale and type, but was run very close by Mr. John Evens'

C

Burton Missy by Professor (200), a typical milking cow. Mr. J. Langham's Brandon Nonpareil, who was second in 1902, was reserve. The judges placed at the top of the two-year-old heifers Binfield Brilliant by Conisholme Samson (2164), out of Binfield Jewel by Dowsby Cambridge Duke 13th (355), the property of Sir Robert Wilmot, Bart., Binfield Grove, Bracknell; the next place being occupied by Foston Gem 2nd by Dowsby Waterloo Oliver 2nd (C.H.B. 72,372) from Foston Gem by Stenigot A. (965). This last was shown by Mr. J. T. Cox, Foston Lodge, Leicester. The third prize went to a heifer by Saltfleet Actor (1664) from a Benniworth Kelstern (28) cow, shown by Mr. Tom Bett, Benniworth; and Mr. J. Marriott's Cropwell Laura, and Mr. J. Langham's Brandon Gem were respectively reserve and highly commended. All the heifers in this class were a beautifully topped lot. Mr. Bett showed a similarly bred heifer in the yearling class, taking premier honours; and Mr. J. Marriott's Cropwell Pride 2nd by Birthday (1782) out of Cropwell Pride by Red Knight (924), was placed second, with a heifer by Ormsby Baronet 2nd (2283), shown by Messrs. J. W. Farrow & Sons, Strubby Manor, in the third place. Mr. J. Langham's Brandon Ruby, was reserve.

In the Shorthorn classes at the Lincolnshire Show, Mr. John Evens was reserve number for the best cow over four years old, in-milk or in-calf, with Burton Missy; but in a class of eleven for the best Shorthorn dairy cow or heifer, registered or accepted for registry in Coates' Herd Book or the Lincolnshire Red Shorthorn Herd Book, he swept all before him, winning the three money prizes and the reserve with Ruby 4th, C. Star 2nd, Nancy and Profitable 2nd.

In the Lincoln Red classes the first prize for bulls two years old or upwards went to Mr. E. H. Cartwright for Benniworth 4th (629) by Keddington (151), out of a cow by Benniworth Kelstern (28); he was a bull of great scale and quality and very even in flesh. Mr. Everett King's Northboro' Cromwell 4th (2587) had again to be content with second place, and Mr. J. Marriott's Birthday (1721) with third honours. But Cropwell Sovereign (2485) again held pride of place among the bulls under two years of age, having now grown into a strong, well-developed bull. A short-legged, shapely bull of nice quality took the second prize in Horkstow Diamond (2919) by Bumper 2nd (1793) out of Horkstow Expectation by Prince Imperial (C.H.B. 69,299), the property of the Hon. G. B. Portman, and bred by Mr. W. B. Swallow, Horkstow. Mr. W. Chatterton's Nero (2991), who had been placed third in the

class for bull calves the year before at Boston, occupied a
similar position at Lincoln, and Mr. T. A. B. Walker, Crow
Park, Notts., was reserved with Keal Chieftain (2934). The
bull calves were led by Mr. J. Langham's Brandon Xmas Gift
(3256) by Brandon Lord Chancellor (2121) out of a cow by
Collingwood ; Mr. G. Freir, Deeping St. Nicholas, being second
with Deeping 28th by Western Sea King (2070) out of a Luton
Chief (447) dam ; and Mr. W. S. Fox, Potterhanworth, third
with a son of Field-Cornet (2515) and a cow by Burtevan (2122),
Captain E. M. Grantham's Keal Duke by Conbalcom (1831).
was reserve. Mr. E. H. Cartwright's Miss Calceby 3rd, a cow
of beautiful type and quality that had been awarded the third
prize the year before in class for cows in milk, was now ad-
vanced to the top of the class, having behind her Norbury
Duchess by Poolham Duke 2nd (886), shown by Mr. S. B.
Carnley, Norbury House, Alford, and Mr. John Evens' Burton
C. Star II. by Professor (200). Red Peony by Tattershall
Grange No. 1. (984), the property of Mr. C. Hensman, Fulletby,
Horncastle, was reserve. The Special Prize for the best
Lincolnshire Shorthorn Red milk cow went to Mr. Evens for
Burton Ruby 4th by Professor (200), and Burton C. Star II.,
from the same herd, was reserved. The judges awarded first
prize for heifers under two and a half years old to Mr. T. Bett's
heifer by Saltfleet Actor (1664) out of a cow by Benniworth
Kelstern (28) ; the second and third prizes going to Mr. J. T.
Cox, Foston Lodge, Leicester, for Red Rust 3rd by Dowsby
Waterloo Oliver II. (C.H.B. 72,372), out of a Kirmond (160)
dam, and Foston Gem 2nd by the same sire, but from Foston
Gem by Stenigot A. (965). Mr. J. Marriott's Cropwell Laura
was reserve. Mr. Bett and Mr. Cox also carried off the first
and second prizes in the class for heifers under eighteen months
old, the former with an heifer similarly bred to the winner in
the previous class, and the latter with Foston Brandon Buttercup III. by Foston Duke (2204), from Foston Brandon Buttercup by Chancellor (332). This last was undoubtedly the best
in the class, but was seriously amiss on the day of the show.
Mr. J. Marriott's Cropwell Pride 2nd by Birthday (1782), was
third, and the reserve number went to a heifer by Ormsby
Baronet 2nd (2283) shown by Messrs. J. W. Farrow & Sons.
The Silver Bowl for the best pair of milch cows went to Mr. J.
Evens for Burton Ruby 4th and Burton Nancy ; Burton C.
Star II. and Burton Profitable, from the same herd, being
reserved.

Mr. Evens' Ruby Spot by Professor (200), who was fourth in the milking trials at Tring, stood one place higher in the inspection class there ; and further honours went to the Burton Herd at the London Show, where Jessie V. was third, and at the Lincoln Autumn and Christmas Shows, where Missy was first and second respectively.

1904. At the second exhibition of the Royal Agricultural Society, at Park Royal, in *1904*, six classes were provided for the Lincoln Reds.

Here Mr. Everett King's Northborough Cromwell 4th (2587) still further distinguished himself by winning in the old bull class, next to him coming Messrs. R. & R. Chatterton with Head-Porter (2909) by Nonsuch (1292), out of Countess 2nd by Red Prince (211), bred by Mr. W. Chatterton at Hallington. The winner had grown into a massive bull, on short legs, with a good fore-end, while the Stenigot exhibit was of great scale and character, with a good middle. Stenigot Soldier (2655) by Sirdar (1676) out of Stenigot Daisy 5th by Red Prince (211), shown by Mr. John Byron, Normanby-le-Wold, was third. In the class for two-year-old bulls, Mr. W. J. Atkinson's Weston Monar ch2nd (3144), second among the yearlings the previous year, improved his position, being very good in his lines and even in flesh, while Mr. J. Langham's Brandon Christmas Gift (3256), the winner at the County Show in 1903, had to be content with second honours. Mr. J. Marriott's Cropwell Red Earl (2851) was third. Bulls calved in 1903 were headed by Mr. J. Evens' Scampton Expansion (4093), bred by Mr. G. E. Sandars, Scampton ; a lengthy, nice-topped bull, by Keddington Ruby (1243), from a dam by Great Tom of Lincoln (392), Mr. T. Bett's Benniworth 29th, a compactly-built son of Red Chief (2611), came next ; the third prize went to Messrs. R. & R. Chatterton's Stenigot Beau 2nd (4107) by Wolseley (1436), out of Stenigot Belle by Comet (79) ; and Sharpshooter (4099) by Field-Cornet (2515), belonging to Mr. W. S. Fox, Potterhanworth, was reserve.

Brandon Satellite, the winner at the County Show in 1902, won again at Park Royal, in the cow class. She showed a good udder and was lengthy and straight in her lines, and was followed by two animals of good dairy type in S.S. 1st by Saltfleet Sentinal (945) shown by Mr. J. W. Measures, Dunsby, Bourne, and Mr. W. J. Atkinson's Saltfleet Favourite by Grandad (1561). Mr. J. Searby's Stenigot Violet 2nd, such a great prize-winner in her day, was reserve. Mr. T. Bett's yearling winner of the year before, now won in the two-year-old heifer class, with Mr. J. Marriott's Cropwell Pride 2nd next in order, both nice heifers ; and the third prize was awarded to Mr. J. Langham's Brandon Ruby, by Brandon Lord Chancellor (2121), while Captain E. M. Grantham's Keal Polly was reserved. The yearling heifers were led by Kirkby Nonpareil, a smart youngster by Benniworth 4th (629), from Nonpareil 2nd by Ludford (172), shown by Mr. John Todd, Kirkby Green. Then came Healing Dorothy by Healing Champion (2910), out of Healing Tip Top, by Limber Tom (829), the property of Captain the Hon. G. B. Portman, Healing Manor, with Mr. E. H. Cartwright's Keddington Vanity, by Vanguard (2691), a very massive, heavy-fleshed bull by Benniworth IV. (629), from a Hallington (135) cow, out of Keddington Lassie, next, while Mr. J. Marriott's Cropwell Gleam 4th was reserve.

At the Grimsby exhibition of the Lincolnshire Agricultural Society, a strong class of old bulls was led by the Royal winner, Northborough Cromwell 4th (2587), and Head-Porter (2909) ; but the third place was occupied by Mr. E. H. Cartwright's Vanguard (2691), and Normanby Beggar Chief (1963), by Beggarman (1457), the property of Mr. G. J. Browne, Tothby, was reserve. Brandon Christmas Gift (3256) came to the front for Mr. J. Langham, in the class for bulls under two years of age, improving on his position at Willesden, and beating the Royal winner Scampton Expansion (4093), who, however, was still placed in front of Benniworth 29th (3215). Strubby Red Coat M. (3632), by Ormsby Baronet 2nd (22⁹3 , belonging to Mr. B. Simons, The Grange, Willoughby, Alford, was reserve.

The best bull calf under a year old, shown by Mr. W. J. Atkinson, was a son of Saltfleet Combination (4075), who was by Saltfleet Bonus (3582), the calf being out of that fine cow, Saltfleet Favourite, by Grandad (1561). The next place was occupied by Cropwell Banister (3309), by Bonfire (C.H.B. 80,524), from a cow by Stenigot (248), the property of Mr. John Marriott, and the third ribbon went to Mr. J. Langham's Brandon Champion (3768), by Brandon Lord Chief Justice (2425), from a Dowsby Lincoln Waterloo 9th (718) cow. The Monk (3656), by Bigby Jumbo (2750), the property of Mr. W. B. Sleightholme, Grayingham Manor, was reserved. The special prize for the best bull shown by a tenant farmer went to Brandon Christmas Gift.

Norbury Duchess improved upon her position of the previous year and took premier honours in the cow class for Mr. S. B. Carnley; the second rosette being awarded to Mr. J. W. Measures for a cow by Saltfleet Sentinel (945); and the third to Mr. W. J. Atkinson's Saltfleet Favourite, by Grandad (1561), the latter being the dam of the winning bull calf, and each a cow of excellent type and dairy character. Mr. J. Marriott's Cropwell Sunshine by Lincoln Sailor (1597) was reserve. The leading heifer under two-and-a-half years of age proved to be Mr. T. B. Freshney's Ruby 12th, by Benniworth 4th (629), from Ruby 2nd by Saltfleet Eclipse (227), a grand type of a Lincoln Red Shorthorn, of great scale, very even in flesh, and a good colour. A cow by Saltfleet Actor (1664), belonging to Mr. T. Bett, and Mr. J. Marriott's Cropwell Pride 2nd, by Birthday (1782), took the other two prizes; Captain E. M. Grantham's Keal Polly, by Conbalcom (1831), was reserve. Mr. J. Todd's Kirkby Nonpareil, by Benniworth 4th (629), from Nonpareil 2nd by Ludford (172), was considered to be the best yearling heifer, next to whom the judges placed Surfleet Dorothy, by Scampton Bloodsucker (3040), belonging to Mr. A. Smith, Surfleet, and Mr. J. Marriott's Cropwell Gleam 4th, by Birthday (1782). Mr. J. Langham's Brandon May Queen, by Brandon Chief Justice (2125), was reserve. Mr. Freshney's beautiful heifer Ruby 12th won the two special prizes for cows; and in the Shorthorn Dairy Class open to both breeds of Shorthorns, Mr. J. Evens carried off the three money prizes and the reserve card in a class of twelve, his representatives being Burton Creamy, by Burton Scamp (2448), Burton Cross 2nd, by Professor (200), Burton C. Star 2nd, by the same sire, and Burton Rose 4th, by Red Rover (C.H.B. 77,619). The latter had previously taken the third prize for Dairy Cows at Newark. In the open Shorthorn classes Mr. J. Langham's Brandon May Queen was second to Mr. H. Dudding's Lady Emma 3rd, in a class for heifer calves, and Mr. Langham's Brandon Christmas Gift (3256), and Mr. John Evens' Scampton Expansion (4093) were respectively second and third in a class of 13 bulls under two years of age, shown by members of the Lincolnshire Agricultural Society, the winner being Mr. H. Dudding's Fencote.

Besides taking first and second prizes in the Shorthorn Butter Test at the London Dairy Show with Fleet II. and Cross II., the former a great prize-winner, and like the latter by Professor (200), Mr. J. Evens also secured a third prize with Burton Milkman, while Ruby Spot was fourth, both in the Milking Trials, Butter Test and Inspection.

1905. At the Royal Show, in *1905*, the old bull class was headed by Mr. T. B. Freshney's Saltfleet Bonus (3582), by Red Monarch (C.H.B. 77,605), from a cow by Lord Knightley (170), who was by Cambridge Duke 29th (C.H.B. 60,440). He was bred by Mr. Riggall, Well, Alford, and proved a most successful sire, as well as a great prize-winner. The second prize went to Mr. E. H. Cartwright's Stenigot Bloom Boy (3611), by Red Chief (2611), out of Stenigot Bloom 4th, by Wolseley (1436), bred by Messrs. R. & R. Chatterton, Stenigot ; and the third to Mr. S. Crawley, Hemington, Oundle, for Weston Monarch 2nd (3144), by Red Monarch (C.H.B. 7765), out of Weston Charm, by Pippin's Pride (C.H.B. 71,157), whose registered prefix to his name denotes his birthplace. Sharp-shooter (4099), by Field-Cornet (2575), from a Burtevan (2122) dam, belonging to Mr. James Cartwright, Dunston Pillar, Lincoln, led the two-year-old bulls ; and the next two places were occupied by Captain Portman's Healing Blucher (3913), by Benniworth Actor (2105), and Mr. Langham's Brandon Champion (3761), by Brandon Lord Chief Justice (2425). Both prizes went to Messrs. R. & R. Chatterton for yearling bulls, they being taken by Stenigot Gwynne and Stenigot Primate, both by Red Chief (2611).

The cows in milk were a good lot, and the winner (who was also the Bath and West winner), was Mr. J. Evens' Saltfleet Favourite, whose breeder, Mr. T. B. Freshney carried off the second prize with the beautiful heifer Ruby 12th, Norbury Duchess, a previous winner, taking the third prize, but Mr. J. Langham's Brandon Satellite, the winner in this class the previous year, was no higher than reserve. Premier honours for two-year-old heifers fell to Mr. J. Todd's Kirkby Nonpareil who won at the Royal as a yearling ; and next in order came Mr. W. J. Atkinson's Ruby 14th, by Regent (C.H.B. 73,398), and Captain E. M. Grantham's Keal Nancy, by Conbalcom (1831). Brandon Red Rose scored for Mr. J. Langham in the yearling class, a pretty heifer by Stroxton Victor (C.H.B. 82,420), behind whom the judge put Mr. J. Marriott's Cropwell Belle II., by Scampton Drama (3598), and Messrs. R. & R. Chatterton's Stenigot Daisy 14th, by Head-Porter (2909), while Mr. W. J. Atkinson's Red Rosette 6th was reserved. Both of the Milk Yield Prizes went to Mr. J. Evens for Burton Primrose 2nd, by Knight of Chewton (C.H.B. 68,867) and Saltfleet Favourite.

At the County Show, at Grantham, the Royal winner maintained his position among the old bulls, but the other two prize-winners changed places, and Sharpshooter, the two-year-old winner at Park Royal, stood reserved. In the class for bulls between one and two years of age Stenigot Gwynne (4995) scored for Messrs. Chatterton. He was by Red Chief (2611), out of Stenigot Gwynne III., by Commander (80), and was afterwards sold to go to Uruguay. Next to him stood Surfleet Baronet (4127), by Surfleet Baron (2666), belonging to the Exors. of the late Mr. A. Smith, Surfleet; and a son of Red Chief (2611), and Stenigot Red Daisy carried off the third prize for Messrs. R. &. R. Chatterton. Mr. S. Crawley's Royal Standard (3557), by Prince Louis (C.H.B. 81,921), was reserve. The pick of the bull calves was Mr. G. J. Brown's home-bred Tothby Virtuoso 3rd, by Kirkby Virtuoso (2947), out of a Baron Ormsby 2nd (25) cow; and below him stood a son of Brandon Lord Chief Justice (2425), and Brandon Nonpareil; and a calf by Under-Porter (3126), from Strubby Favourite was third, these respectively the property of Mr. J. Langham and Messrs. J. W. Farrow & Sons. Saltfleet Bonus was declared the Champion Lincoln Red bull, for which honour Western Monarch II. stood reserved.

Mr. E. H. Cartwright's handsome cow, Star of the Night, by Shooting Star (1674) claimed pride of place among the cows in milk or in calf; Ruby 12th taking the second honours as at Park Royal; with Kirkby Nonpareil third; while Saltfleet Favourite, first at the Royal and second at Peterborough, got no higher than reserve. Captain Grantham's Keal Nancy stood at the top of the heifers between the ages of one year and two-and-a-half years; the second and third prizes and the reserve card going respectively to Mr. W. J. Atkinson's Ruby 12th, Mr. J. Todd's Kirkby Nonpareil, and Mr. J. Langham's Brandon Sunrise. Red Rosette 6th, belonging to Mr. W. J. Atkinson, was placed first among heifers from 12 to 18 months old; next coming Messrs. J. W. Farrow & Sons' Strubby Pride, and Mr. J. Marriott's Cropwell Belle 2nd; and Messrs. R. & R. Chatterton's Stenigot Daisy 14th, in the reserve place. Mr. J. Langham's Brandon Lovely, by Brandon Blend (3250), and Mr. S. B. Carnley's Norbury Duchess 3rd, by Lord Augustine (C.H.B. 81,472), took the two prizes for heifer calves. The Female Championship went to Keal Nancy, with Star of the Night as runner-up.

The chief incident of interest in connection with the Lincolnshire Show this year was the meeting of Mr. T. B.

Freshney's Saltfleet Bonus (3582), and Mr. Philo Mills' wonderful bull King Christian of Denmark (C.H.B. 86,316), for the championship of the show-yard. No fewer than seven judges were called in to decide between their respective merits, and amid a scene of the utmost excitement the coveted rosette was handed to the Lincoln Red. He was a bull of great scale, was very even in flesh and well ribbed up, and had a good back. King Christian of Denmark took many first and championship prizes all over the country, and at the dispersal sale, at Ruddington, in *1906*, he was sold to Mr. A. W. Hickling, Adbolton, Notts., for 900 guineas. All the first prizes in the classes for Shorthorns, open to farmers of not more than 500 acres, fell to Lincoln Reds, and in the dairy class open to all pure-bred Shorthorns, the three money prizes and reserve number were taken by Mr. J. Evens with the Burton cows, Rose and Quality 3rd, who had been placed third at Newark, Dairymaid 5th, and Saltfleet Favourite.

At Peterborough, Mr. E. H. Cartwright's Star of the Night was placed in front of Mr. J. Evens' Saltfleet Favourite ; and Mr. J. Todd's Kirkby Nonpareil and Mr. W. J. Atkinson's Ruby 14th, were respectively first and second in their class. Mr. Evens' Dairymaid V. and Creamy were third and fifth in a class of 21 dairy cows. At the London Dairy Show the Burton herd took second and fourth prizes with the cows Cross II. and Fleet II. ; first for three-year-old heifers with Rose V., by Red Rover (C.H.B. 77,618), and second in the same class with Violet II.

1906. At the first revival of the Royal Agricultural Society's migratory shows, held at Derby in *1906*, eight classes were placed at the disposal of Lincoln Reds, including that for the Milk Yield Prizes. At the County Show, at Gainsborough, there were seven classes for the breed and two championships. The old bull class at the Royal was headed by Mr. Robert Chatterton's Stenigot Regal Chief (4998), by Red Chief (2661), from Stenigot Flower 4th by County Member (83), a bull of excellent Shorthorn character, compact and nicely put together, with a characteristic head. Mr. S Crawley's Western Monarch 2nd (3144) was second, and Mr Anselm Parker, Walsall, took the reserve ticket with Saltfleet Bonus (3582), the winner of the previous year, who was shown in very poor condition. Surfleet Baronet (4127), a second prize-winner at the County Show the year before, carried off the first prize in the two-year-old bull class for Mr. C. T. Garland, Moreton Morrell, Warwick ;

the second going to Mr. Robert Chatterton's Stenigot Royal Chieftain (4597), by Red Chief (2611), out of Stenigot Flower 4th ; while Mr. W. J. Atkinson's Western Monarch 4th (4187), by Royal Crest (C.H.B. 82,149), was reserve. Mr. J. Langham's Brandon Grenadier (4274) who took a second prize as a calf at the County Show the previous year, led the yearling bulls ; Benniworth 39th (4239), by Saltfleet Echo (3038) carrying off the second prize for the Exors. of the late Mr. T. Bett ; with Mr. R. Chatterton's Gosberton Beau (4391), by Clan Mac-donald (C.H.B. 78,597), and Mr. J. W. Measures' Grainthorpe Red 12th (4552) respectively third and reserved. At the top of the class for cows in milk, calved before 1902, stood Mr. J. Evens' Burton Quality 3rd, by Red Rover (C.H.B. 77,618), from Burton Quality by Collingwood (C.H.B. 57,074), a cow destined to play a very important part in the show ring ; and the second prize was awarded to Burton Primrose 2nd, who took the premier honours in the milk yield classes at the Royal in 1905. Captain E. M. Grantham's Keal Nancy again carried off a first prize by winning in the three-year-old heifer class, with Mr. C. T. Garland's Surfleet Dorothy, by Scampton Bloodsucker (3043) next ; and Brandon Lovely, by Brandon Blend (3250), belonging to Mr. J. Langham, headed the two-year-old heifers, followed by Mr. C. T. Garland's Strubby Pride, by Lord Augustine (C H.B. 81,472), and Favourite, by Saltfleet Bonus (3582), shown by Earl Egerton of Tatton, Tatton Park, Knutsford, with Mr. Chatterton's Stenigot Gwynne 2nd, as reserve number. But Stenigot Buttercup scored for the latter breeder in the yearling heifer class, a pretty heifer by Red Chief (261), from Stenigot Buttercup 10th, by Wolseley (1436) ; the second ribbon going to Captain Grantham's Keal Hilda, by Scampton Excavator (4084), and the reserve position was occupied by Lord Egerton of Tatton's Beauty. Both prizes in the Milk Yield class went to Mr. J. Evens, for Burton 4th, by Longfellow (1605), and Burton Primrose 2nd, by Knight of Chewton (C.H.B. 68,867).

At the Royal Counties milking trials, Mr. Evens was first both in the milk test and in the class for the best dairy cow, with Iris ; while at Peterborough the Royal winner, Quality III., was not only considered the best Lincolnshire Red cow, but the pick of the 16 dairy cows.

At the County Show, at Gainsborough, Saltfleet Bonus (3582) made amends for his defeat at the Royal by heading the old bull class, Western Monarch 2nd (3144) having again to be content with second place ; while Surfleet Baronet (4127), first in the two-year class at Derby (R.A.S.E.), was third. Horkstow Diamond (2919), by Bumper 2nd (1793), the property of Captain the Hon. G. B. Portman, Healing Manor, was reserve.   Bulls between the age of one and two years were led by Mr. Robert Chatterton's Gosberton Beau (4391), third at the Royal, and now placed above the winner there.   Mr. J. Langham's Brandon Grenadier (4274), and the second prize-winner, Benniworth 39th (4259) shown by the Exors. of the late Mr. T. Bett.   Mr. E. H. Cartwright's Keddington Baron (4881) by Saltfleet Bonus (3582), was reserve.   Captain E. M. Grantham's Keal Tommy (4880), by Scampton Excavator (4084), was considered to be the best bull calf ; and next in order the judges placed Donington Baron (4355), by Surfleet Baron (2666), the property of Mr. J. Drewery, Donington, Spalding.   Mr. S. B. Carnley's Norbury Draco (4486), and Mr. R. J. Epton's Northolme Charter were respectively third and reserve.

Saltfleet Bonus (3582) was again proclaimed the Champion Lincoln Red bull in the Show, and once more Western Monarch 2nd (3144), was runner-up.

Miss Calceby 3rd, the winner of the class for cows in milk or in calf at the Lincolnshire Show, at Lincoln, in *1903*, and now eight years old, again repeated her fine performance, beating that good three-year-old heifer, Captain Grantham's Keal Nancy, and the Royal winner, Mr. J. Evens' Burton Quality 3rd.   Mr. E. H. Cartwright's Keddington Skipworth 5th by Benniworth 4th (629), was reserve.   The Royal winner, Mr. J. Langham's Brandon Lovely, scored again in the class for heifers over one and under two-and-a-half years old.   Mr. E. H. Cartwright's Twilight and Mr. R. Chatterton's Stenigot Buttercup 16th—the first by Vangaurd (269), and the second by Red Chief (2611)—coming next in order ; with Mr. E. H. Cartwright's Keddington Favourite 4th, also by Vanguard, reserve.   Another Royal winner came to the front in the class for heifers over one and under 18 months old, in Mr. Robert Chatterton's Stenigot Buttercup 16th, and Captain Grantham's Keal Hilda was again second ; but Lord Egerton of Tatton's Beauty had to stand below Mr. B. Simons' Trusthorpe Fairy, by Buscot Loyalty (C.H.B. 80,591) and a reserve heifer by Norbury Cato (2993) shown by Mr. G. Laughton, Belchford. But the latter breeder headed the heifer calves with a daughter of Norbury Cato and a cow by Saltfleet Ruby Chief (944) ; the second and third prizes going to Mr. E. H. Cartwright's Keddington Vain Girl, by Stenigot Bloom Boy (3611), and a heifer by Sir James (C.H.B. 82,335), belonging to Mr. G. Marris, Kirmington House, Brocklesby, and the reserve card to Captain Grantham's Keal Nancy 2nd.   Miss Calceby 3rd carried off the female championship, for which Brandon Lovely stood reserved.

The majority of the prizes in the classes for Shorthorns of either breed, shown by members or Lincolnshire tenant farmers, were won by Lincoln Reds.

At Peterborough, Mr. W. J. Atkinson's Western Monarch 4th (3144) and Mr. R. Chatterton's Gosberton Beau (4391), each took a first prize ; Mr. John Evens' Burton Quality III. carried off two first prizes, the second to her in the cow class being Clara, shown by Messrs. Fletcher & Andrews, Horninghold, Uppingham ; and Mr. J. Langham's Brandon Lovely and Mr. R. Chatterton's Stenigot Buttercup 16th were both prize-winners.

At the Eastern Counties Show, Mr. Evens took first for the best dairy cow, with Burton Jessie 5th, by Red Rover (C.H.B. 77,618) ; and at the London Dairy Show, besides numerous victories in the milk and butter tests, he obtained second and fourth prizes for heifers under three years of age, with Cowslip IV. and Rose VII. Whitefoot VII. also gained him a first prize at the Lincoln Fat Stock Show.

1907.   In *1907* for the first time, prizes were offered by the Association for bulls entered for the annual sales conducted under their auspices at Lincoln. There were four prizes in each of two classes, the first being for bulls calved between March 1st, 1905, and March 1st, 1906, and, at the head of a strong lot of 18, the judges had no difficulty in placing Scampton Hermes (4972), bred and owned by Mr. G. E. Sandars, Scampton, Lincoln, a beautifully level and neat bull by Keddington Ruby (1243), from a cow by Digby Conqueror (2182), who came of Great Tom of Lincoln (392) and King Hal (156) descent. This bull was afterwards sold to Mr. W. B. Swallow, Wootton Lawn. Ulceby, at 140gs. The Exors. of the late Mr. T. B. Freshney took the second prize with Saltfleet Dragoon, by Saltfleet Bonus (3582), out of a Nonsuch (1292) cow, who was later on knocked down to Mr. G. J. Brown, Tothby, at 62 guineas. Digby Herald 1st (4801) stood third, by Horkstow Herald (3919) and the property of Mr. P. F. Brown, Digby Manor, Lincoln, who eventually disposed of him to Mr. H. Caudwell, Midville, at 46gs. ; and the fourth ribbon went to Mr. Reuben

Roberts, for Thimbleby Curly Coat, by Scampton Excelsior
(4085), purchased by Mr. W. R. Sharpe, Swineshead, at 62gs.
The other class was for bulls calved on or after March 1st, 1906,
and the first two prizes went to Messrs. T. & W. Dickinson,
Worlaby, for Bonby Excursionist 4th, and Bonby Excursionist
5th, both by Scampton Excursionist (4019), and out of cows by
Revolving Light (1658). The first of these was sold to Mr. J.
G. Williams, Tring, at 65gs., and the other to Mr. W. Thorn-
alley, Ashby, at 46gs. Mr. G. Langdale, Market Stainton,
stood next with Volunteer, by Anderby Red Coat (4217), for
whom Messrs. Bowthorpe & Co., London, paid 43gs.; and the
fourth prize went to Kirmington 1st, by Red Cap 2nd (3546),
the property of Mr. G. Marris, Kirmington House, Brocklesby,
and sold to Mr. L. W. Stephenson, South Thoresby, at 36gs.

With the Royal Agricultural Society holding its exhibition
at their county town, the Lincolnshire Red Shorthorn Association made special efforts to make the display of their cattle a
worthy one, and visitors from all parts of the globe were
greatly struck with their wealth of lean flesh and great milking
capabilities, and indeed they surpassed all other breeds for
general excellence. Four prizes were given in each of the
twelve classes, and in addition there were two champion prizes
and two special prizes confined to members of the Lincolnshire
Agricultural Society. In the class for cows in milk, Mr. J. G.
Williams, Tring, won with Nonpareil 9th, by Conisholme Boy
(347), a wide deep cow, bred by Mr. E. H. Cartwright; the
second prize giong to Mr. J. Todd, Kirkby Green, for Kirkby
Nonpareil, by Benniworth IV. (629). The third place was
occupied by Keddington Pearl, by Bigby (319), shown by Mr.
G. Marris, Kirmington House, Brocklesby, and the fourth by
Clara, by The Bard (986), the property of Messrs. Fletcher &
Andrews, Horninghold, Uppingham. Mr. F. Scorer's Bracebridge 90th, was reserve. The cows with calf at foot were led
by Mr. J. G. Williams, Keddington Skipworth 5th, another big
square cow by Benniworth IV. (629), bred by Mr. E, H. Cartwright; next to whom stood Mr. Robert Chatterton's Cropwell Laura, by Lincoln Sailor (1597), bred by Mr. J. Marriott,
and Messrs. J. W. Farrow & Sons' cow by Red Curly Coat (923),
while Mr. J. Evens' Burton Scarlet, by Scarlet Cloth (5101),
was fourth, and Mr. J. F. Rawnsley's Cherry Ripe, by Croft
Bilsby (2169), was reserve. The pick of the heifers in milk,
calved in 1904, was undoubtedly Mr. George Marris' Keddington Favourite, by Vanguard (2691), a sweet heifer with good
top and bottom lines, and very wide and deep. The second

prize went to another Vanguard heifer of Mr. E. H. Cartwright's breeding in Twilight, belonging to Colonel C. A. Swan, C.M.G., Sausthorpe Hall. Saltfleet Poppy, by Benniworth 4th, (629), the property of the Exors. of the late Mr. T. B. Freshney, and Mr. J. Langham's Brandon Red Rose, by Stroxton Victor (C.H.B. 82,420) stood third and fourth, and Lapwater A., by Poolham Scampton (3013) shown by Messrs. C. Hensman & Son, Fulletby Grange, was reserve. Heifers calved in 1905, were headed by Mr. W. J. Atkinson's Weston Pink, by Clan Macdonald (C.H.B. 78,597), very pretty and square and of good quality ; the second ribbon going to Otby Glitters 2nd, by Dunsby Sentinel (1535), the best in the class save for a badly-set-on tail ; the third to the Exors. of the late Mr. G. Laughton, Belchford, for a daughter of Norbury Cato (2993) ; and the fourth to Trusthorpe Fairy, by Buscot Loyalty (C.H.B. 80,591), exhibited by Mr. B. Simons, Willoughby Grange, while Mr. J. G. Williams' Keddington Skipworth 7th, by Saltfleet Bonus (3582), from the winning cow with calf at heel, was reserve. The yearling heifers numbered 24, the bulk of the animals being of excellent quality, so that the judges were a long time in placing them for decoration. But the winner deserved her pride of place, for she is very neat in form, as square as a brick and a nice colour. Ruby Queen by name, she is by Saltfleet Bonus (3582), out of Ruby 12th, which made 105gs. at the late Mr. T. B. Freshney's sale, and she was bred by her owner, Mr. E. Bourne, Louth. The second place was occupied by Messrs. J. W. Farrow & Sons' Cardiff 3rd, by UnderPorter (3126) ; Deeping Princess, by Scampton Exclusion (4088), belonging to Mr. G. Freir, Deeping St. Nicholas, was third ; Saltfleet Blooming, by Stenigot Bloom Boy (3611), shown by the Exors. of the late Mr. T. B. Freshney, stood fourth ; and Mr. J. Langham's Brandon Ruby Nonpareil, by Brandon Christmas Gift (3256) was awarded the reserve ticket. The first three prizes for cows in milk went to Mr. J. Evens, for Burton Ruddy 5th, by Red Rover (C.H.B. 77,618), Spot 5th by the same sire, and Violet by Stamp End (247) ; while the fourth prize was awarded Miss K. Carleton for Burton Young Cherry, by Butter Boy (1477), and Mr. Evens' Cork 3rd, another cow by Red Rover, was reserve. It was the first time out for the first and second prize-winners, and both were of rare dairy type. All the four money prizes for heifers in milk, calved in 1904, went to the Burton herd for Margaret 3rd by Saltfleet Daring (2620), Prophetess 5th, by Burton Rex (2131), Marjorie 3rd, by the same sire, and Ruddy 6th, by Saltfleet Darling. The special prize for the best cow or heifer shown by members of the Lincolnshire Agricultural Society, went to Mr. E. Bourne's Ruby Queen, for which Mr. G. Marris' Keddington Favourite 4th, stood reserved ; while the championship of the females was awarded to Mr. J. G. Williams' Keddington Skipworth 5th, with Ruby Queen as runner-up. The milk yield prizes will be referred to elsewhere.

The old bull class was headed by Mr. J. Langham's Brandon Blend (3250), by Brandon Lord Chancellor (2121), and bred by the exhibitor, quite the squarest and best-fleshed in the class. Mr. Robert Chatterton's Under-Porter (3126), by Saltfleet Ascent (1671) stood next followed by Weston Monarch 2nd (3144), a bull by Red Monarch (C.H.B. 77,605), shown by Mr. S. Crawley, Hemington, Oundle, while the fourth rosette was handed to the veteran, Keddington Ruby (1243), then eleven years old, and somewhat light in condition, but making a wonderful show for his age. Mr. J. Drakes' Grange Captain (3897), by Upp Hall (2368), was reserve. At the top of the class for bulls calved in 1903 or 1904 the judges placed Scampton Exile, the property of Mr. B. Rowland, Ivy House, Wainfleet, and bred by Mr. G. E. Sandars, Scampton, being by Keddington Ruby (4092), from a King Hal (156) dam. He was a massive, heavy-fleshed bull, and was built on right lines. Later on he was awarded championship honours, which also fell to him again in 1906. The next place was occupied by Mr. W. J. Atkinson's Western Monarch 4th (4187), by Royal Crest (C.H.B. 82,149) ; Mr. J. Searby's Croft Aubourne (4325), by Calceby Marvel (2453), was third ; Surfleet Baronet (4127), by Surfleet Baron (2666), the property of Mr. C. T. Garland, Moreton Morrell, Warwick, occupied the fourth place ; and Mr. R. Chatterton's Hagnaby Admiral received the reserve ribbon. Easily the first among the two-year-old bulls stood Mr. G. E. Sandars' Brandon Grenadier (4274) quite of the right type, of beautiful quality, and good to touch. He won at the Royal Show the previous year, and was bred by Mr. J. Langham, being by Brandon Lord Chief Justice (2425), from Brandon Nonpareil, by Chancellor (332). Then followed Mr. R. Chatterton's Stenigot Violet Chief, by Red Chief (2611), a massive bull, with good top and bottom lines ; Donington Baron (4355), by Surfleet Baron (2666), belonging to Messrs. J. Drewery & Son, Donington ; and Breedon Champion, a son of Saltfleet Bonus, shown by Mr. H. W. Blunt, Ashby-de-la-Zouch ; while Captain E. M. Grantham's Keal Blois (4873), by Scampton Excavator (4584) was handed the reserve card. Mr.

R. Chatterton's very promising yearling Stenigot Duchess Beau, by Stenigot Beau 2nd (4107) carried off premier honours in the young bull class ; the second prize going to Mr. J. Evens' Burton Hermit 2nd, by Burton Hermit (3783) ; the third to Mr. J. Langham's Brandon Grenadier 2nd, by Brandon Majestic 2nd (4276) ; while the fourth and reserve tickets fell to Harpswell Baron (4857), by Horkstow Herald (3919), the property of Mr. T. Turner, Harpswell, and to Mr. J. G. Williams' Bonby Excursionist 4th, by Scampton Excursionist (4089). Both the special prize for members of the Lincolnshire Agricultural Society and the championship for bulls went to Mr. B. Rowland, for Scampton Exile (4092), who was run very close by Mr. G. E. Sandars' Brandon Grenadier (4274) to whom some would have given pride of place.   There was, of course, no County Show that year, but a number of classes were reserved for purely Lincolnshire exhibits, on the fourth day of the show.   But the pick of the Lincoln Reds, being Royal prize-takers, were debarred from competing, and so the C.H.B. Shorthorns, which had not been so successful in the open classes, had matters all their own way.   It was an absurd arrangement, and served no useful purpose whatever.   But in a strong class for shorthorn cows or heifers, in milk or in calf, Mr. J. Evens' Burton Qaulity 3rd, who was not eligible for the Royal classes, was a good winner, beating Mr. E. Bourne's Lincoln Red Ruby 12th, Captain E. M. Grantham's Keal Nancy, and Mr. H. Dudding's C.H.B. Royal Lily.   The winner was of great scale and quality and of undeniable milking type, and by beating the champion Lincoln Red females of 1904, 1905 and 1906, in the above class, and twice beating the 1907 champion at Peterborough and Tring, she proved herself to be without doubt the premier Lincoln Red of the year.   Her pedigree has been given elsewhere.

Next to Mr. Williams' cow, at Peterborough, came two more Burton cows, Scarlet and Daisy, as third and fourth prize-winners ; and the yearling bull Hermit II. (4724), by Burton Hermit (3783), out of that famous cow Saltfleet Favourite, was second in his class.   Besides, the victory of Quality III., and the triumph in the milking trials, mentioned elsewhere. Mr. Evens' Margaret III., by Saltfleet Darling (2620), stood second in the three-year-old heifer class ; and at the Derbyshire Show he took a first prize for bulls with Hermit II., and a second for Quality III. in the class for cows.   Hermit II. was also awarded first and special for bulls at the London Dairy Show, where Dolly II., by Carlton Red (330), was second for dairy cows of any breed or cross ; Co-Fox IV. and Violet VI., were third for pair of dairy cows, any breed or cross ; Ruby Spot II. and Nancy III. were first and second for Lincoln Red dairy cows ; and C. Star VII. was first in the open class for heifers under three years of age ; these besides numerous victories in milk and butter tests. Then at Lord Tredegar's Show at Newport, Burton Jessie V was second for dairy cows ; and at the Lincoln Autumn Show, Mr. Evens secured first prizes with Harpswell Laughter (4858), by Horkstow Herald (3919), in the class for bulls under 14 months old, and Fuchsia III. in the cow class.

1908. Mr. Evens opened the *1908* show season with a first at Newark, with Burton Quality III. ; also carrying off a great number of prizes in the milk and butter tests at the Oxfordshire and Bath and West Shows.

For the second year in succession a show was held on April 23rd, 1908, in connection with the Lincolnshire Red Shorthorn Association's sale of bulls at Lincoln, four prizes being again offered in two classes, the first for bulls calved between August 1st, 1906 and March 1st, 1907, and the other for bulls calved on or after March 1st, 1907. Mr. G. E. Sandars again headed the older class with a son of Keddington Ruby (1243), this being Scampton Invader (5609), who was own brother to the highest-priced bull at the 1905 sales, and to the cow that won first prize at the Lincoln Fat Stock Show. He was a very heavy-fleshed bull, good in his lines, and a beautiful colour ; and he was the only son of old Keddington Ruby on the ground, the bulls. by which impressive sire, numbering 40, had averaged over 60gs. at these sales. Invader was bought by Mr. Granville Sharpe, Baumber Park, at 97gs. The second prize in this class went to Mr. F. Scorer's Brace-bridge Baron 2nd (5172), by Welbourne Red Baron (3693), for whom Mr. Reuben Roberts subsequently paid 68gs. Mr. Roberts himself carried off the third ribbon with Thimbleby Emperor, by Scampton Excelsior (4085) who was knocked down to Messrs. T. & W. Dickinson at 76gs., and the fourth went to Messrs. Dickinson's Bonby Baron, by Keddington Baron (4881).

The younger bulls were led by Saltfleet Friar (5603), by Saltfleet Dragon, a very neat youngster afterwards sold to Mr. F. B. Wilkinson, Edwinstowe, Newark, at 54gs. ; while next to him stood Deepden Benniworth (5290), by Bingham Lodge Benniworth (2752), the property of Mr. E. Ashley, Godman-

D

chester, Hunts., for whom Mr. John Evens subsequently paid 54gs. on account of his belonging to a famous milking family. The third and fourth prizes went to Mr. B. Simons' Willoughby Tom (5752), by Kirkby Tom (4464), and Messrs. Dickinson's Bonby Excursionist 20th, another promising young bull by Scampton Excursionist (4089).

At the Royal Show at Newcastle, the Lincoln Reds were of course much fewer in numbers than they were at Lincoln the previous year, but then there was a special effort on the part of breeders to make the display a notable one on the occasion of the Royal Society's visit to their County Town. But there are always some who are anxious to obtain Royal honours, and so included in the seven classes were many animals of high merit. In the old bull class, where five paraded, the winner turned up in Mr. W. J. Atkinson's Western Monarch 4th (4187), who last year took second place to the animal that was afterwards awarded the championship of the breed. He showed a lot of quality, and shorthorn character, was short on the leg, and even in flesh, and he fully deserved his place. Mr. R. Chatterton's Hallington Neptune (3904), the second prize-taker was a massive and heavy-fleshed bull, but did not walk so well, nor had he as good shoulders as the winner. Mr. R. Chatterton's Cicero (4314) and Mr. H. W. Blunt's Breedon Champion were respectively third and reserve. Mr. John Evens showed the best two-year-old bull in Burton Hermit 2nd (4724), very square and of nice quality. His sire, Burton Hermit (3783), was out of that famous winner at milking trials, Burton Fleet II., and he himself was out of Saltfleet Favourite, a noted cow, and a winner of first prizes both at the Royal and Lincolnshire Shows. Mr. R. Chatterton's Flower Chief Consul (5344), by Red Chief (2611), was the only other exhibit in the class. But there was little difficulty in picking out Mr. Robert Chatterton's Stenigot Bloom Boy 2nd (6359) as the winning yearling, for he was built on the most approved lines, and had a typical shorthorn head. Mr. J. G. Williams' Pendley Boxer, by Somercotes Bonus (4577) came next in order; Mr. John Evens' Deepden Benniworth (5290) stood third, and Mr. J. G. Williams' Pendley Count was the reserve number. The pick of the cows in milk calved in or before 1904 was undoubtedly Mr. George Marris' handsome Keddington Favourite 4th, a wealthy heifer of great scale, with a good top and bottom line, and a capital udder. She comes of Mr. E. H. Cartwright's breeding, and was a winner in her class at the Royal last year. Both Mr. J. G. Williams and Earl Egerton of Tatton showed nice

cows, in Starlight, by The Count (1396), and Enderby Lass, by
Lord Chancellor (1606). Mr. John Evens' Burton Quality V.,
by Burton Peer (2797) took the reserve rosette. Mr. Crawley
chose Captain E. M. Grantham's Keal Hilda for chief decoration among the three-year-old heifers, and Messrs. C. Hensman
& Sons' Fulletby Beauty A. to lead the two-year-olds. She is
very lengthy, and has a good back, but is not so deep and wide.
The other prizes went to Messrs. C. Hensman & Sons' Fulletby
Royal, and Mr. J. Evens' Burton Royal Maid III., in the one
class, and to Mr. J. G. Williams' Pendley Pearl, and Messrs.
C. Hensman & Sons' Fulletby Pansy III. in the other. The
yearling heifers were a good lot, but although the winner was
both long, wide, and short on the leg, she was by no means the
best topped one of the class. Pendley Skipworth by name,
she belonged to Mr. J. G. Williams, and was a daughter of
Keddington Baron (4881). Perhaps the best back in the
bunch belonged to Messrs. Hensman's Fulletby Duchess,
though she wants to grow down a bit to be a good one;
she was placed below Captain Grantham's Keal Barbara and
Mr. John Searby's Croft Heroine 2nd, the first by Keal Blois
(4873), and the other by Croft Aubourne (4325).

The Lincoln Reds made a brave show, as they always do
at the County Exhibition, at Sleaford, and the spectators round
the ring were much struck with their great scale and wealth,
and their undoubted dairy qualities. Mr. B. Rowland's five-
year-old bull, Scampton Exile (4092), repeated his last year's
performance at the Royal by winning both first for old bulls
and the championship of the Lincoln Reds, and he was also
reserved for the championship of the Show, being second only
to the mighty Chiddingstone Malcolm, champion Shorthorn
of the year. In his class he beat this year's Royal and
Peterborough winner. Mr. W. T. Atkinson's Weston Monarch
4th (4187), who stood reserved for the Lincoln Red championship. Scampton Exile was by that most impressive sire
Keddington Ruby (1243), and he was bred by Mr. G. E.
Sandars, Scampton. Mr. R. Chatterton's Hallington Neptune
(3904), and Mr. J. H. Brown's Scampton For Ever (4558) came
next in order. There was nothing to beat the Royal winner
two-year-old bull, Stenigot Bloom Boy 2nd (6359), shown by
Mr. R. Chatterton. He was followed by Mr. J. G. Williams'
Pendley Right Stamp, Mr. F. B. Wilkinson's Saltfleet Friar
(5603), and Mr. J. Evens' Deepden Benniworth (5290), the
latter being reserve number. Mr. S. Crawley's calf, Crimson
King, a youngster of great promise who won in his class,

began the series of victories with which is name is associated. By Duke of Barrington 70th (C.H.B. 94,959), he was out of Enderby Princess V., by Saltfleet Combatant (2319) ; and he beat Mr. G. J. Brown's youngster, by Calceby Express (4302) ; Mr. J. Searby's Croft Cayhurst, by Kirkby Virtuoso (2942), and Mr. F. B. Wilkinson's Sherwood Speculation, by Queen's Bitrhday (4511). The cow class furnished a truly wonderful display of Lincolnshire Red Shorthorn matrons, such scale and substance, and such milking capabilities with quality withal, impressing the onlookers to no small extent. But there was no mistake in putting Mr. John Evens' splendid cow, Burton Quality 3rd, at the top of the class. The previous year she beat the champions for that and the two preceding years at Peterborough, and beat the Royal winner at Tring, and she won at Peterborough again this year. She was bred at Burton and is by Red Rover (C.H.B. 77,618), Mr. E. Bourne's Ruby 12th of the late Mr. T. B. Freshney's breeding was second, and last year's Royal winner, Mr. J. G. Williams' Keddington Skipworth 5th, third, while Saltfleet Poppy, shown by the Exors. of the late Mr. T B. Freshney, stood reserved. Messrs. C. Hensman & Sons led the younger heifer class with Fulletby Peony 3rd, quite a nice one, but the judges had to call in Mr. Robert Fisher, Leconfield, to place the Royal winner, Mr. J. G. Williams' Pendley Skipworth, in front of Mr. J. Tomlinson's daughter of Northholme Astonishment in the class for heifers between 12 or 18 months old. Mr. J. Searby's heifer calf beat Mr. Williams' Peterborough winner. Burton Quality 3rd won the championship, for which Ruby 12th stood reserved. The prizes in the County classes, in which the prize-winners in the open classes are debarred from competing, were about equally divided between Lincoln Reds and C.H.B. Shorthorns, Mrs. Webb & Sons, Messrs. S. E. Dean & Sons, and Messrs. H. Dudding, J. Evens, C. Hensman & Sons, J. Searby, and G. Marris being the most successful. In the dairy class open to both breeds of Shorthorns, the County cattle had it all their own way, Mr. John Evens taking the premier rosette, the third and the reserve (the latter with last year's Royal winner), while Mr. F. Scorer was second.

At Peterborough, Weston Monarch 4th (4187) was put in front of Burton Hermit II. (4724), with Mr. E. Ashley's King Cowslip (5455) in the third place. In the younger bull class, Mr. J. G. Williams' thick well-fleshed youngster, Pendley Right Stamp, was the winner, and he was followed by Mr. J.

Measures' Dunsby Red II., with Messrs. C. Hensman & Sons' Willoughby Athlete, third. The Burton cows Quality III. and Spotted V. stood first and third in their class, divided by Mr. J. G. Williams' Keddington Skipworth. Weston Pink and Fulletby Beauty A. took first and second prizes for Messrs. Hensman in the Heifer class, beating the Royal winner Pendley Skipworth; but in the class for heifer calves Mr. Williams won with a daughter of Keddington Comet (3443), Messrs. Hensman standing second and third with heifers by Scampton Formula (4562). Besides winning in various dairy trials, Mr. John Evens took a first at Newark and a second at Nottingham with Burton Quality III. as well as a third at the London Dairy Show, where he also took first and third awards for Lincoln Red dairy cows with Spotted V. and Profit III. Here too he stood first and second with Ruby XII. and Plenty V., in the open heifer class, under 3 years of age.

1909. The 14th Bull Sale of the Lincolnshire Red Shorthorn Association, at Lincoln, was quite a success, in that a greater number of bulls were disposed of than has been the case for very many years, and a fresh average was set up. The year before 187 bulls were sold at an average of £28 5s. 8d., and in 1907, 210 were disposed of at £26 8s. 5d. apiece, while the previous year 166 went to average £27 10s. 9d. This year, of the 258 lots entered in the catalogue, nine were not forward, 11 were passed unsold through the ring, and 238 changed hands at an average cost of £28 6s. 6d., the total sum realized being £6,742 1s. 6d. Taken altogether, it was a much better display of the breed than I have ever seen before at these sales; and each year finds them shorter on the leg, and with better backs, while still maintaining their substance and wealth of lean flesh. The two classes at the Show before the sale were quite good ones, and included some animals that will be heard of again. The winner in the older class was Mr. J. W. Measures' Dunsby Red 3rd (6017), a beautiful quality bull, and with a capital top and bottom line. He is by Sausthorpe Red 12th (1496), from a cow by Commander A. (4552), and he was afterwards awarded the Championship. When he eventually entered the sale ring he was the object of considerable competition, Mr. F. B. Wilkinson, Edwinstowe, Mr. J. Evens, Burton, and Mr. T. Wallis, Lincoln, being chiefly concerned, and after the latter had retired, he was knocked down to Mr. Wilkinson at 165gs.

Next to him in his class and reserve for Championship honours stood Scampton Jupiter (6332), the property of Mr

G. E. Sandars, Scampton, who has so many times obtained both the highest price of the sale and the best average. This is a very massive, heavy-fleshed bull, by rare old Keddington Ruby (1243), from a Great Tom of Lincoln (392) dam, though with not perhaps the top line of his conqueror, nor is he quite so neat behind the shoulders. Mr. T. Wallis bid well for him, but he eventually fell to Mr. L. Hobbs, of London, purchasing for abroad, at 90gs. The third in the older class was from the same herd, Scampton Judas (6325) by name, and very even in his lines, with a lot of substance, and Mr. T. Wallis secured him at 90gs. This bull is by Mr. Sandars' 200-guinea purchase Brandon Grenadier (4274), and he is out of a Keddington Ruby (1243) cow. The fourth prize in this class was taken by Mr. Henry Kirke's Ravendale 3rd (6272). afterwards sold to Mr. R. J. Epton, Wainfleet, at 54gs. The judges (Mr. Robert Wright, Nocton, and Mr. W. J. Atkinson, Weston St. Mary) had no trouble in picking out Mr. Sandars' Scampton Justinian (6335) for the first prize in the younger class, and a most promising youngster he is, as square as a brick and good to touch. He is similarly bred to Scampton Judas (6325) ; and the first crop of calves by Brandon Grenadier certainly point to the fact that the price paid for him was not a dear one. Justinian fell to the bid of Mr. John Searby, Croft, at 100gs. Next to him in this class stood Anderby Pilot (5793), the property of Messrs. J. W. Robinson & Son, for whom Mr. Campion, North Thoresby, paid 50gs. The third prize went to Mr. John Byron's Normanby Solus (6219), who was knocked down to Messrs. Allerdice & Co., Liverpool, at 80gs., while the fourth prize-winner, Mr. Sandars' Scampton Julius (6330) made 85gs., to Mr. A. Turner, Oadby, Leicester.

Considering the distance from their native county the Lincolnshire Red Shorthorns made a very creditable display at the Royal Agricultural Society's Exhibition at Gloucester. Only two old bulls paraded, and no mistake was made in awarding the first prize to Mr. Percy Hensman's Scampton Exile (4092), still showing a bull of great scale and character and very even in flesh. For the past two years, when the property of Mr. B. Rowland, he was champion Lincoln Red bull, and the year before he was reserved to Sir R. P. Cooper's celebrated bull, Chiddingstone Malcolm for the Championship of the Lincolnshire Show. Mr. R. Chatterton's Hallington Neptune (3904) who was second at the Royal and third at the County Show last year, took the other prize. Mr. J. Searby's Tothby Gem 4th (5703), by Tothby Scotchman (3673) led the two-year-old bulls, and the other two prizes went to Mr. A. P. Brandt, Bletchingley Castle, Surrey, and Mr. R. Chatterton, for King Louis (5457) and West Ashby Wolseley (5741) respectively, Mr. F. B. Wilkinson's Dunsby Red Third (6017), the winner in the older bull class at the Lincolnshire Red Shorthorn Association's Show and Sale at Lincoln in April, was absolutely undecorated.

Mr. Robert Chatterton's Stenigot Duke, by Stenigot Duchess Gwynne (5633), was somewhat lucky to beat Mr. J. Searby's Scampton Justinian (6335) in the yearling bull class, for the latter, who is by Brandon Grenadier (4274) and won in the junior class at Lincoln in April is much the neater and truer to type. Mr. A. F. Nalder showed an improving youngster in Redlam Power, who was awarded third prize, and Mr. J. Evens' Pendley Count (6251) was reserve.

The first prize in a good class of cows was awarded to Mr. J. G. Williams, Pendley Manor, Tring, for Keddington Skipworth 5th, a rare type of a Lincoln Red matron. She was bred by Mr. E. H. Cartwright, Keddington Grange, and is by Benniworth 4th (629), out of Keddington Skipworth 3rd, by Bigby (319), and was placed second at Peterborough, and third at the Lincolnshire Show last year, while in 1907 she was the champion Lincoln Red cow at the Lincolnshire Show. Mr. Williams also took second prize with Benniworth Bloom, by Saltfleet Actor (1664) and Kenl Hilda, who was a first prize-winner at the Royal last year, took the third ribbon for Captain E. M. Grantham, the reserve number being Mr. B. Wilkinson's Donington Crawley.

Mr. Williams also won in the three-year-old heifer class with Pendley Pearl, a Benniworth bred one, by Saltfleet Echo (3038), out of Benniworth Pearl, by Saltfleet Actor (1664), a big square heifer that was reserve number at Tring last year ; the second prize going to Mr. F. B. Wilkinson for Stenigot Queen 9th, by Stenigot Bloom Boy (3611). A third first prize fell to Mr. Williams for the two-year-old heifer Pendley Skipworth, who won first prizes both at the Royal and Lincolnshire Shows last year, and took a second at Tring and a third at Peterborough. She is out of the winning cow Keddington Skipworth 5th, being by Keddington Baron (4881). Next to her stood Mr. Percy Hensman's Fulletby Marvel, by Croft Aubourne (4325), and a third prize fell to Mr. J. Tomlinson's Birthorpe Belle, who was second at the County Show, at Sleaford. Captain E. M. Grantham's Keal Barbara, who was second at the Royal and third at Sleaford, had to be content with a reserve number.

A good class of yearlings was headed by another from Mr. Williams' famous herd at Tring, this being Pendley Starlight 2nd, by Keddington Comet (3443), from Starlight, by The Count (1356), a short-legged, compact little lady, with a good top and bottom line. Mr. C. F. Bett stood second with Benniworth Pink, pretty and neat and good in colour, by Somercotes Bonus (4577), out of Benniworth Actor (2015) cow, and the third and reserve ribbons went to Mr. F. B. Wilkinson's Sherwood Bonnie Girl and Mr. A. F. Nalder's Redlan Queen 1st, the latter a very improving heifer.

There was quite a good display of the breed, when the Lincolnshire Agricultural Society held its 39th show at Louth, in 1909. Six very good bulls paraded in the class for two-year olds and upwards, in the Lincoln Red ring, and premier honours went to the Royal winner, Mr. Percy Hensman's Scampton Exile (4092). This grand old bull, now six years old, was champion of the Lincoln Reds in 1907 and 1908, and was reserved to Sir R. P. Cooper's grand roan bull, Chiddingstone Malcolm, for the championship of the Lincolnshire Show in 1908. This year he was not shown in quite such condition, and this undoubtedly lost him the championship. Mr. Robert Chatterton's Hallington Neptune (3904), who was second to him at the Royal, again stood next to him. Mr. W. J. Atkinson's Weston Monarch 4th (4187) and Mr. J. Searby's Tothby Gem 4th (5703), stood next in order. The class for bulls between one and two years of age was a very strong one, but there was no doubt about the winner, for Mr. S. Crawley's Crimson King (5258) stood out among the 16, although such an excellent lot. He won as a calf at the Lincolnshire Show last year, and this year was champion at the Northamptonshire Show at Thrapston, beating all comers, including the Irish Champion. But the second and third were distinctly lucky to be where they were, and the reserve number and at least two others might have been put in front of them. Mr. A. F. Nalder's Redlan Power and Mr. F. B. Wilkinson's Dunsby Red 3rd (6017), were the other prize-takers, Mr. E. Abraham's Otby George (6240) being the reserve number. In the class for bull calves, after Mr. Robert Chatterton's Stenigot Freedom had been selected for premier honours, one of the Shorthorn judges had to be called in to decide between the respective merits of Mr. E. Bourne's Ruby King and Lord Fitzwilliam's Wentworth Earl, the better back of the former giving it the preference. Crimson King secured the championship of the breed for which Scampton Exile was runner-up.

In the class for cows in milk or in calf there was nothing to touch Mr. John Evens' Burton Quality III., who had beaten every rival during the last four years, and had lost nothing of her great merit as a typical Lincoln Red matron. Another good sort stood second in Mr. E. Bourne's Ruby 12th, one of the late Mr. T. B. Freshney's old tribe, with a fine lot of cows behind them, including Mr. F. B. Wilkinson's Somercotes Starlight and Mr. A. P. Brandt's Stenigot Bloom 10th. Mr. J. G. Williams' Pendley Skipworth, too, was easily first in her class, but Mr. P. Hensman's Fulletby Marvel II. and Captain E. M. Grantham's Keal Barbara were of good merit. In the class for under two-years-old heifers, Mr. Williams took pride of place with Pendley Starlight II., also of Keddington blood. Both were beautiful heifers. Mr. E. Bourne's Ruby Queenie and Mr. C. F. Bett's Benniworth Pink came next in order. Captain Grantham's Keal Nancy III. led the heifer calves, a very promising youngster, and behind her came Mr. J. G. Williams' Pendley Rosebud and Mr. G. Freir's Deeping Daisy. But it was a terrible disappointment to see Burton Quality III. beaten for the female championship by Pendley Skipworth, and it is really difficult to see where the judges saw the superiority of the younger heifer over this grand old cow, whose great scale and magnificent udder alone entitled her to the honour. However, she took the championship for tenant farmers or members farming or residing in the County as a small consolation for a big robbery. All three prizes and the reserve ticket for Dairy Cows went to Mr. John Evens, each of the money-takers being out of cows that have won at milking trials.

At Peterborough, Scampton Exile (4092) scored another victory and again took pride of place over Hallington Neptune (3904) and Saltfleet Friar (5603) in the older bull class ; while Mr. Robert Chatterton's Stenigot Duke repeated his Gloucester victory, Mr. J. G. Williams with Pendley Corrector, and Mr. S. Crawley with Laddie being the other prize-takers. Burton Quality III. was easily first in her class, behind her being Mr. F. B. Wilkinson's Sherwood Starlight and Mr. E. King's Red Rose II. In the class for heifers under three years of age Mr. Williams took first of place with Pendley Skipworth ; Fulletby Marvel II. and Birthorpe Belle coming next in order as at the Royal ; and the Pendley Manor herd scored again with the heifer calf Pendley Rosebud, the other prizes going to Mr. F. B. Wilkinson's Sherwood Special and Mr. G. Freir's daughter of Stenigot Duchess Gwynne (5633).

Mr. John Evens took a second prize at the Rutland Show with Burton Ruby Spot III., and at the London Dairy Show, stood first, second and third in the Lincoln Red dairy class with Quality V., Spotted V. and Ruby XII. ; second and fourth in the Open heifer class with Pride V. and Cross IV., fourth for pair of dairy cows with Ruby Spot II. and Bramble-finch, and fourth with Fanny in the class for single dairy cows.

1910. The 15th Show and Sale of Lincoln Red bulls, held under the auspices of the Lincolnshire Red Shorthorn Association, took place at Lincoln, on April 28th, when out of a total entry of 313 bulls 295 were brought forward, 277 finding purchasers at an average price of £25 5s. 7d. This was very much below the figure of recent years, and must be set down to the fact that the supply was slightly in excess of the demand.

The judges were Messrs. W. J. Atkinson and Robert Wright, and there were twenty-eight entries in the class for Bulls calved between August 1st, 1908, and March 1st, 1909, and forty in that for Bulls calved on or after March 1st, 1909, and in both classes the display was highly creditable to the breed. The awards in the Senior class were as follows :— First prize, Mr. George Marris, Kirmington Forester 13th (6952) ; second, Mr. W. H. Smyth's Elkington Marshman 1st (6778) ; third, Mr. E. Abraham's Otby Hercules (7125) ; fourth, Mr. G. E. Sandars' Scampton King of Trumps (7038) ; reserve, Messrs. J. N. Robinson & Son's Anderby Neptune (6537). In the Junior class they were :—First and second prizes, Mr. G. E. Sandars' Scampton King of the Valley (7123) and Scampton King of Hearts (7121) ; third, Mr. Percy Hens-man's Fulletby Friend (6803) ; fourth, Captain C. L. Prior's Strubby Boy 6th (6160) ; reserve, Mr. Reuben Roberts' Thimbleby Commander (7187).

The official judges disagreed as to the respective merits of Mr. George Marris' Kirmington Forester 13th (6952) and Mr. G. E. Sandars' Scampton King of the Valley (7123) for the championship, and Mr. C. W. Tindall was called in to decide the question, giving his verdict in favour of the older bull, who was perhaps a little neater on the top of the tail. The champion, who was calved on January 5th, 1909, was bred by Mr. Marris at Kirmington, and is a beautiful type of Lincoln Red, full of quality, yet possessing the size and substance which has made the breed so famous and which it is so essential should not be lost while striving for greater neatness of form. He combines the blood of two famous families, for his sire, Scampton Forester (4557) was bred by Mr. G. E. Sandars, and his dam, Keddington Pearl, a prize-winner at the County Show and the Royal, was by Bigby (319) and bred by Mr. E. H. Cartwright at Keddington Grange, near Louth. Scampton King of the Valley (7123), the runner-up for championship honours, is one of the handsomest bulls Mr. Sandars has ever bred. He is by that impressive sire Brandon Grenadier (4274), a Royal winner, for whom Mr. Sandars paid 200gs. at Alford Fair in 1906 ; and he is from a Great Tom of Lincoln (392) dam, being own brother to Scampton Invader (5609), a first prize-winner at the Association's show in 1908, and is out of the same cow as Scampton Fortress (4563), the highest-priced bull in Lincoln Fair in 1905.

Although the classes at the Royal Show were not so well filled as at the annual County Show, the entries for 1910 compared very favourably with those of other years, while the quality was distinctly superior to anything that has been previously shown at the Royal, with the exception of the Lincolnshire Exhibition of *1907*. This year the entries numbered forty-three as compared with fifty-two last year, and never have the cattle been so well brought out and so uniform in type. In the old bull class the first prize went to Mr. J. G. Williams, of Pendley Manor, Tring, for Grange Prince (4843), who was bred by Mr. E. S. Cartwright, at Keddington Grange, Louth. This bull is of nice quality with good thighs and top line, lengthy and low. Except that he does not carry his head very well, there is no fault whatever to find with him. Mr. Robert Chatterton took second place with a weighty, short-legged bull, Red Chief III. (4939), by Red Chief (2611) out of Withern Stenigot ; while the third place was occupied by King Louis (5457), second at the Royal last year. He was bred by Mr. S. Crawley, Hemington, Oundle, and was shown by Mr. Augustus P. Brandt, of Bletchingley Castle, Surrey. He is a deep, square bull, and much improved since last year. Mr. F. B. Wilkinson, of Cavendish Lodge, Edwinstowe, stood reserve with Saltfleet Friar (5603), reserve at Gloucester last year, and third at Peterborough. Only two bulls were forward in the class for two-year-olds, the first prize going to Stenigot Comet, the property of Mr. Robert Chatterton, Welbourne Hall, Lincoln. He is by Keddington Comet II. (7144), out of Stenigot Choice III. He has beautiful quality, is of nice type with a good top and bottom line, and moves well. Captain the Hon. G. B. Portman, Healing Manor, Lincolnshire, took the second prize with Wentworth Earl (7248), bred by Earl

Fitzwilliam at Wentworth. This bull took third prize at
the Lincolnshire Show last year, and was the highest-priced
bull at the Lincolnshire Fair in April. He is wide and deep,
and has plenty of size and substance. First for yearling bulls
went to Mr. John Evens, Burton, Lincoln, with Kirmington
Forester XIII. (6952), bred by Mr. George Marris, of Kirmington, Brocklesby. He is by Scampton Forester (4557), out of
Keddington Pearl, was first in the old bull class at the Association's show at Lincoln, and was afterwards awarded the
championship. He is of great size and substance, and a good
colour, with a good top and bottom line. The second prize
went to Mr. F. B. Wilkinson for Scampton King of the Valley
(7123), by Brandon Grenadier (4274), who won in the young
bull class at the Association's Show at Lincoln, and was reserve
for the championship, and is a lengthy, wide bull. Mr. Augustus
P. Brandt took the third prize with Bletchingley Brennus
(6595), by Keddington Comet (3443), and the reserve was Mr.
Henry Neesham's Canwick Ruby III. The class for cows in-
milk was an exceptionally good one, and the great size and
substance and milking qualities were most favourably commented on around the ring. Mr. J. G. Williams won with
Pendley Pearl, as he did last year. Square and level, good to
look at in front and behind, short on the leg, a capital colour,
and showing a good bag, she is by Saltfleet Echo (3038) from
Benniworth Pearl, and was bred by the late Mr. T. Bett. Mr.
Percy Hensman, of Fulletby Grange, Horncastle, took second
prize with Keal Hilda, third at Gloucester. She is a deep,
roomy cow, with a good back, and is by Scampton Excavator
(4084) out of Keal Red Daisy. Third place was occupied by
Donington Crawley, shown by Mr. F. B. Wilkinson, a lengthy,
good cow, reserve at the Royal last year, and sired by Lord
Chancellor 10th (1900). The reserve ribbon went to Mr.
Augustus P. Brandt, for Stenigot Bloom X., reserve at the
Lincolnshire Show last year, a cow of sweet type, with considerable dairy qualities. Two beautiful heifers stood first
and second in the class for three-year-olds, both shown by Mr.
J. G. Williams. The first prize went to the Royal and Lincolnshire champion of last year, who also took a first prize at Peterborough, Pendley Skipworth, bred at Tring, by Keddington
Baron (4881) out of Keddington Skipworth V., herself a Royal
winner. She is a perfect Lincolnshire Red heifer of great scale
and substance, and very symmetrical. The second prize-
winner, Pendley Starlight, was also bred at Tring. She is by
Keddington Baron (4881), is well grown, deep and lengthy,
and has a good back. Next to her comes Mr. Percy Hensman's

Fulletby Tindall II., of nice quality, but not quite furnished in her thighs. Mr. Hensman also took the reserve card with Fulletby Marvel II. First and second prize again fell to Mr. Williams in the class for two-year-old heifers, the first being last year's Royal and Lincolnshire winner, Pendley Starlight II., by Keddington Comet (3443), out of Starlight, of great width and depth, and very wealthy in flesh. The other heifer, Pendley Countess, is also well grown and lengthy, and is also by Keddington Comet (3443). The third prize-winner was shown by Mr. F. B. Wilkinson, Benniworth Pink by name, second at the Royal last year and third at the Lincolnshire. This heifer has a beautiful back, and very wide, well-sprung ribs. Mr. Brandt's Deeping Daisy III., third at the Lincolnshire Show last year, stood reserve. No mistake was made in placing Mr. Brandt's Bletchingley Bellona at the head of the yearling heifers, for she is of good colour and type, and very neat and pretty. She was bred by Mr. George Freir at Deeping St. Nicholas, and is by Buscot Rupert. Mr. Williams took second place with Pendley Violet IV., by Bonby Excursionist IV. (5161); Captain E. M. Grantham, of West Keal, Spilsby, was third with Keal Doris; and Mr. John Evens reserve with Burton Fancy, who might have gone up a place.

The County Show was held at Spalding, in *1910*, and a good class of old bulls paraded, Mr. P. Hensman's well-known Scampton Exile (4092), wonderfully level and active for a seven-year-old, led the way, with Mr. Williams' Grange Prince (4843), first at Liverpool, in the second place, and Mr A. P. Brandt's King Louis (5457), third. The reserve was Mr. C. E. Scorer's Kirkby Imperial (4896), a straight, level bull.

In the yearling class the first three were again as at Peterborough, Mr. John Evens' Kirmington Forester 13th (6952), which has never looked back since winning first prize at the Lincoln Show and Sale, being still at the head of his class. Mr. F. B. Wilkinson's Scampton King of the Valley 2nd (7123), was second, and Mr. A. P. Brandt's Bletchingley Brennus (6595) third. Mr. W. Thornalley had reserve with Bonby Excursionist (5162), a useful bull.

Bull calves were quite a good though a somewhat mixed class. Mr. J. Measures took first prize with Dunsby Red 6th, a bull just a week short of a year old. He is wonderfully well grown, and appears likely to finish a big one, but he is straight and level, and handles well. A very smart youngster was second in Mr. E. Abraham's Otby Kinsman. He has not the spread of the winner, but gives nothing away in quality. Captain E. W. Grantham's Keal Rollo was third, nice-bodied bull, with much that is pleasing about him, but he has got quite enough white. Mr. S. Crawley's Crown Prince, which was reserve, might have gone a place higher. The special prize for the best bull went to Mr. Hensman's Scampton Exile (4092), with Mr. Evens' yearling Kirmington Forester 13th (6952) reserve, and the order was the same for the special prize restricted to county competition.

In the cow class the judges had plenty of scope for selection, but first honours did not go past the Peterborough winner, Mr. Williams' Pendley Skipworth. Pendley Pearl, from the same herd, which was second at Peterborough, and first at the Royal, failed to find favour, and this big-framed cow was well down the list with no more than a commendation. The second prize was awarded to Captain Grantham's Keal Nancy, which successfully disposed of another good cow, Keal Hilda, bred in the same herd, and exhibited by Mr. Hensman. The latter was second at the Royal, and there is little to choose between them. The reserve card went to Mr. F. B. Wilkinson's Sherwood Chimes, which thus took the same position as at Peterborough.

In the two-year-old class Mr. Williams was again success-ful with Pendley Starlight 2nd and Pendley Countess, which held similar positions at both Liverpool and Peterborough ; Mr. F. B. Wilkinson's Sherwood Daisy and Benniworth Pink were third and reserve respectively.

Yearlings were quite a fair class, though the quality was not very even. First prize went to Mr. T. H. B. Freshney for Saltfleet Ruby, a well-grown heifer, dark in colour. Mr. G. E. Laughton had second prize with an unnamed heifer of consider-able promise, and a get of Gwynne Cowslip (6059). Mr Brandt's Bletchingley Bellona, which was winner at Liverpool, was now third Still she made an improvement on her Peter-borough position, beating Mr. F. B. Wilkinson's Fulletby Quality 11th, which was third on the previous occasion. There were but four heifer calves, and a smart lot they were. Mr. G. Freir's Deeping Pansy took first place, her growth and development filling the eye. Her rivals however have some disadvantage in age, as Mr. J. G. Williams' Pendley Lassie and Mr. J. Searby's Croft Aubourne Gem, which were second and third, are late autumn calves, but smart young heifers.

The champion prize for the best cow or heifer was won by Mr. Williams with his two-year-old, and his first prize cow was reserve. The special for the best cow or heifer exhibited from the county fell to Capt. Grantham's Keal Nancy, Mr. Freshney having reserve for his first prize yearling. In the class for dairy cows or heifers, either C.H.B. or Lincoln Red, Mr. John Evens took first and second prizes with Burton Bramblefinch and Burton Butterfly. The third prize went to Mr. C. E. Scorer for Bracebridge No. 3b, and Earl Fitzwilliam's Wentworth Rose was reserve. There was a new class for Lincoln Red bulls that had been in regular use, exhibited in ordinary working condition, and not fed up for showing. There was an entry of fifteen, but on the whole it was rather an unsatisfactory class. The difference between ordinary condition and show condition with some Lincoln Red breeders is evidently slight, and well-conditioned bulls had a good start from the others. Mr. Neesham's Waddingworth Gunner 19th (6453) was placed first, Mr. B. Roland's Dunsby Red 2nd (6016) followed, and Mr. C. E. Scorer's Kirkby Imperial (4896) was third.

The challenge cup for the best bull of either breed was won by Mr. F. Miller's Good Friday, and Mr. Hensman's Scampton Exile (4092) was reserve ; many experts were of opinion that the judges had made a mistake in so placing them.

There were several classes of Lincoln Red Shorthorn cattle at Peterborough. The first was for bulls over eighteen months old. Here Mr. P. Hensman exhibited the old champion Scampton Exile (4092), by Keddington Ruby (1243), who was worthy of his position. Next to him was placed Mr. J. G. Williams' Royal winner, Grange Prince (4843), by Stenigot Bloom Boy (3611), exhibiting any amount of substance and a good top though he might be better packed behind the shoulders. Third was Mr. A. P. Brandt's King Louis (5457), by King Counsel (3960), a bull with a good head and plenty of substance. Mr. F. B. Wilkinson's reserve bull, Hallington Neptune (3904), was shown somewhat uneven in fleshing. Bulls under a year and a half were not a very good class, but Mr. J. Evens with his Royal winner, was easily first, Kirmington Forester 13th (6952), being by Scampton Forester (4557), a very level bull. Second to him came Mr. F. B. Wilkinson's Scampton King of the Valley (7123), by Brandon Grenadier (4274), a useful type of bull with a strong outlook. Mr. A. P. Brandt's Bletchingley Brennus (6595), by Keddington Comet (3443) came third, and Mr. Hensman's Fulletby Athlete reserve.

In the cow or heifer class, first and second cards came to

Mr. J. G. Williams for Pendley Skipworth and Pendley Pearl, both of whom are young cows, the former a clear winner in the matter of scale, width, and wealth of flesh. The second is not quite so level in flesh, and the shape of her bag might be better, although she has nice Shorthorn outlook. Mr. F. B. Wilkinson's third, Sherwood Daisy 3rd, by Queen's Birthday (4511), is a cow of good scale, but she is moderate in the middle of the back. Mr. Nalder was reserve with Redlan 23rd, and Mr. Wilkinson h.c. with Sherwood Chimes. Heifers under 2½ years were not numerous. First and second prizes went to Mr. Williams' Pendley Starlight 2nd and Pendley Countess, the former having the most substance. She is a big, wealthy cow by Keddington Comet (3443). Then came Pendley Countess, a little further from the ground than the winner, but still a massive cow, and by the same sire. Third was Mr. F. B. Wilkinson's Fulletby Quality 4th, showing nice Shorthorn characteristics, and reserve came Mr. A. P. Brandt's Bletchingley Bellona.

In the small heifer calf class Mr. Freir had the victory with Deeping Pansy, by Foston Clement (3877), a very nice youngster of shapely proportions. Indeed, there were three very good calves here, Mr. Williams showing the other two.

At Newark, Mr. John Evens was first for cows with Burton Empress, and third for dairy cows at Peterborough, with Burton Blossom; while at the London Dairy Show he took all three prizes for Lincoln Red dairy cows with Quality V., Ruby XII., and Ruby Spot II., and was first, second and third for Lincoln Red dairy heifers with Rose XI., Ruby Spot VII., and Lady Burton III.

THE MILK PAIL.

Although the Lincoln Reds had long been noted for their motherly virtues, and their ability to raise two or more calves besides their own, it was left to Mr. John Evens, of Burton, Lincoln, to demonstrate to the world, by his truly wonderful successes in milking trials, both in England and Ireland, what the breed, by judicious selection, was really capable of. During the last two or three years Mr. Evens' example has been followed by Mr. Fred Scorer, of Bracebridge, Lincoln, and a few more, and if other Lincoln Red dairy herds were established near the big towns, dairymen would find the breed a profitable one, not only as regards the milk pail, but also in the sale ring.

The characteristics of Mr. Evens' herd are :—Milk, size, flesh, constitution and uniform red colour. The custom up to 1885 was the usual one in the district, viz., breeding and rearing ; the steer calves were grown on for beef, the young cows reared their own calves and were sold at about third calf to go into town dairies. But, in 1885, Mr. Evens decided to take advantage of the nearness to Lincoln to start dairying, selecting to commence with the best bagged cows, and adding other deep milking cows from time to time. And ever since Monday, March 23rd, 1885, the morning and evening milk of each cow has been weighed and recorded, and the yearly milk records of the herd have been published. His first success in the Show Ring was at the London Dairy Show, in October, 1887, when Beauty won the first prize in the Shorthorn Milking Trials, and the Lord Mayor's Champion Cup for the best milker in the Show. The Burton herd was registered in the first volume of the Lincolnshire Red Shorthorn Association's Herd Book, in 1895, it having descended to Mr. Evens from his father and grandfather ; it has now obtained the position of premier Lincoln Red dairy herd and one of the best known dairy herds in the world. To all parts of the globe, wherever milk is in demand, there Mr. Evens' bulls and heifers are sent ; and in the Transvaal, Matabeleland, Natal, Cape Colony, Sweden, Canada, South America, Chili, British Columbia, New South Wales and elsewhere, will be found representatives from the famous Burton herd.

Probably no dairy herd has had such successes in Milking Trials as has Mr. Evens', and the account of his show-yard

E

efforts reads as one long unbroken record of success.   In 1899 he was first, second and fourth in the Milking Trials at the Royal Show, Maidstone, with Whitefoot, Old Profit and Bountiful, the first two having also been first and second in the milking trials at the Royal Show, Birmingham, the previous year.   His victories in inspection classes are mentioned in the chapter dealing with the breed in the Show ring, and so no reference will be made here to his Lincolnshire Show and other successes ; but he was third with Bertha in the milking trials, at the London Dairy Show.   In 1900, Ruby and Judy were first and third at the Bath and West milking trials, and the latter was fourth in a class of 34 at the Hertfordshire milking trials at Tring ; and Mr. Evens also took third with Chance in the London Dairy Show Milking Trials.   The next year found Lady Marjorie first in the open and Ruby Spot first in the Tenant-farmers' milking trials at the Oxfordshire Show ; and Quality was first at the Bath and West, while Dolly secured fourth honours at the London Dairy Show milk tests.   Ireland was invaded in 1902, and C. Star III. took a first in the Royal Show milking trials at Dublin, and fourth at Belfast, and she followed this up with second in the Tenant-farmers' Class at the Oxfordshire Show.   Then Millie Fox was third, and Creamy fourth at Dublin ; and the latter, besides winning at Belfast, took the Challenge Cup for the cow gaining most points in the milking trials, any breed or cross.   Next Fleet II. was second in the open milk tests at the Oxfordshire Show, and Ruby Spot first in the milk trials and second in the butter tests at the Bath and West, at Plymouth ; and Margaret was first in the Royal Counties milk trials at Reading, and first among 61 competitors at Tring, breaking the record there with 75lbs. of milk in 24 hours ; besides being second in the butter test, for Tenant-farmers.   Creamy was third in this class, and third in the milking trials.   That year at the seven principal milking trials in England and Ireland, Mr. Evens won a Challenge Cup, five first prizes and one second.   Next year, 1903, saw Fleet II. winning the first prize and the Challenge Cup, at the milking trials both at Dublin and Belfast, Maud being second in each case.   Then Spot II. was first in the open and Vic II. first in the Tenant-farmers' milk tests at the Oxfordshire Show ; Plenty was second at the Bath and West at Bristol, and C. Star II. first at the Royal Counties at Southampton.   The same year Nancy was first out of 39 competitors at Tring, and second in the butter test, though giving a greater weight than any other cow with 3lbs. 1½ozs. in 24 hours ; Ruby Spot, too, was fourth in the milking trials at Tring ; and at the London Dairy Show Margaret was third both in the milk yield, and butter tests, and was reserve for the £50 Gold Cup for the best all-round Dairy Cow in the Show. The eight principal milking trials in England and Ireland this year yielded two challenge Cups and six first prizes.

In 1904 Mr. Evens again captured the first prizes and Challenge Cups at both Dublin and Belfast, the first with Dairy-maid III. and the second with Young Cherry, who had taken third prize at Dublin, where Maud was fourth, and Emerald was third at Belfast. More victories went to the Burton herd at the Oxfordshire milking trials where Spotted III. was first in the open, and Buttercup first in the Tenant-farmers' classes ; and at the Royal Counties Show, at Guildford, where Dairy-maid V. was placed first, and Ruby IV. second in the milking tests. At Tring again, the largest test ever held in England, 45 cows of all breeds competing in their class, Cross II. was second with 71lbs. 12oz. milk in 24 hours, and Fleet II. fourth with 66lbs. 4ozs., the latter being also placed third for the best cow over four years old. First and second prizes among 20 competing cows went to Fleet II. and Cross II. in the Short-horn butter test at the London Dairy Show ; and Ruby Spot was fourth in the same class, being also fourth both for inspection and in the milking trials.

The year 1905 proved a most successful one for the Burton Herd, which now boasted of 30 cows that had won in public Milking Trials, Butter Tests and Dairy Classes, having carried off seven Challenge Cups, six Cups or Special Prizes, 33 first, 16 second and 12 third prizes. First prize and Challenge Cup, at the Royal Dublin Milking Trials went to Mr. John Evens for the fourth consecutive year with Primrose II., and she also took pride of place in the Oxfordshire Milking Trials, open class, Star II. being second, and first in the Tenant-farmers' class. Next, Dolly II. was second in the Royal Counties Milking Class ; and Primrose II. and Saltfleet Favourite were respectively first and second in the Lincoln Red milk test at the Royal. Then came further triumphs at Tring, where Buttercup was first in the milking trials with 71½lbs. milk in 24 hours, and second in the Tenant-farmers' Butter Test with 2lbs. 8½ozs. in 24 hours ; and finally at the London Dairy Show where Cross II. was second in the Shorthorn milking trials and second for inspection and milking combined in the Tenant-farmers' Class ; and Dolly III. and Violet II. were first and second in the milking trials for heifers.

Next year, 1906, Primrose II. and Chance IV. were first and second in the Special milk yield class at the Oxfordshire Show, and the former, besides taking a first in the open class, took the championship for the best dairy cow while the latter was first in the Tenant-farmers' class. This was followed by a second with Cork III. in the cow class, and a first with Josephine IV. in that for heifers at the Somersetshire milking trials, and two firsts with Iris in the Royal Counties milking trials. The Royal Show honours included first and second in the Lincoln Red milk tests with Burton IV. and Primrose II. and these were respectively second and fourth in the open milk test in which 70 cows of all breeds were competing; while Vic II. was fourth out of 70 in the open butter test. At the Bath and West, Burton IV. was second in the milking trials; and at the London Show, Star was fourth in the butter test and fifth in the milking trials; and Cowslip IV. was first and Pride III. third in the milk test for heifers.

The dairy triumphs of the Burton herd opened in 1907, with first and second in the Oxfordshire Milking Trials with Violet (who gave 72½lbs. of milk in 24 hours) and Fuchsia; and they also took first and second places in the Tenant-farmers' class. At the Royal Show, Vic 2nd and Violet were first and second in the Lincoln Red Milk Yields, the third prize going to the Burton-bred Young Cherry, the property of Miss K. Carleton, Gilford Castle, co. Down, and the fourth to Mr. F. Scorer's Bracebridge 117th. In a Special Milk Yield Class open to all breeds Mr. Evens won with Vic 2nd, the second prize going to Mr. W. P. Vosper's South Devon Honesty 3rd, and the third to Mr. J. Evens' Iris. As a comparison of the total points of the leading cows the following table will be of interest :—

Breed.	Cow.	Points.	Awards.
Lincoln Red,	" Vic 2nd "	83·90	1st twice)
South Devon,	" Honesty 3rd "	80·78	1st & 2nd
Lincoln Red,	" Iris "	78·93	3rd
Shorthorn,	" Warwickshire Hettie "	74·12	1st
Lincoln Red,	" Violet "	73·03	2nd
Lincoln Red,	" Young Cherry "	71·25	3rd
Shorthorn,	" May Duchess "	70·82	2nd
Shorthorn,	" Priceless Princess "	70·17	3rd
Guernsey,	" Goodnestone 2nd "	68·48	1st
Ayrshire,	" Lady Flora "	68·00	1st
Lincoln Red,	" Bracebridge 117th "	67·50	4th

In the general average of all breeds the South Devons come first with 72·09 points. But only two of this breed were entered, as compared with eight Lincoln Reds; and if the average of the two best of these is taken, it works out at 78·49 points, which places them easily ahead of all breeds in the Show. The following were the first three breeds in the table of averages; and the averages of the Jersies and Kerries are given for the sake of comparison :—

Breed.	Number.	Milk, lbs. oz.		Fat. per cent.	Total.
South Devons	2	46	12	4·16	72·09
Lincoln Reds	8	51	1	3·35	67·35
Shorthorns ..	9	49	1¾	3·16	62·64
Jersies	5	37	5⅛	4·50	62·07
Kerries	5	35	4	3·42	57·93

At the Royal Show this year Mr. Evens was awarded first prize among 41 competitors for the best-managed farm of over 300 acres in the County of Lincoln. At the South Eastern Counties Show, Iris was second in the milk test with 69 lbs. 14 oz. of milk, and was reserve for the Gold Medal, 36 competing. At Tring, Young Cherry was first in the milking trials with 71⅜ lbs. of milk, and Iris second with 68½ lbs., thus making four firsts, two seconds and two fourths in the largest and most representative milking trials in England in six years; while at the London Dairy Show where classes were allotted to the Lincoln Reds for the first time, Mr. Evens took first and second with Ruby Spot and Nancy III., Mr. F. Scorer being reserve with Bracebridge 3, B. In the butter test Nancy IV. was first with 2 lbs. 8 oz. of butter in the day, and Young Cherry second. Nancy IV. also won in the milk trials for Lincoln Reds, closely run by Bracebridge 3, B; and the former was awarded the Lord Mayor's Champion Cup for the cow gaining most points in the milking trials of all breeds, Bracebridge 3, B. being reserve, a great triumph for the breed. In 1907-1908, Bracebridge 3, B. was 441 days in milk and gave 1,386 gallons of milk, while in 1908-1909 she gave 1,346 gallons in 301 days. Between June 12th and November 18th, 1910, she gave 976 gallons of milk.

The Burton successes in 1908 again opened at the Oxfordshire Show milking trials, where Cork III. was first both in the open class for any breed, and in the Tenant-farmers' Class, and Fuchsia III. was second in each. There are two classes at the Oxfordshire milking trials, one open to all breeds or crosses, and the other confined to Tenant-farmers' cattle ; Mr. Evens has now carried off the first prize in each class for the last six consecutive years. At the Bath and West, at Dorchester, Tozzle was first in the milking trials for cows, 900lbs. live weight and over, with 66·02 points, beating Mr. Vosper's South Devon Magpie (60·25 points), and Dairymaid 4th (59·67 points).

Then Milker took a second prize at the Royal Counties Milking Trials, at Southampton, and at the Royal Show, at Newcastle, Mr. Evens took first and second prizes in the Lincoln Red Milk Test for the fourth year in succession. Ruby Spot and Cork III. were the two winners above mentioned, and Lord Egerton of Tatton's Enderby Lass 4th was third, with Burton Quality V. reserve. In the milk test open to all breeds and crosses, the Burton herd took second place with Milker, and fifth with Tozzle, Milker also taking the second special prize for most butter in the Show, while in the open butter test she came second, with Ruby Spot and Cork III. fifth and sixth. At Tring Milking Trials, Milker had to be content with third place, but only one tablespoonful of milk divided the first three. In the Lincoln Red Milk Test at the London Dairy Show, Nancy IV. was second and Ruddy V. third, the winner being Mr. F. Scorer's Bracebridge No. 102, while Burton Nancy V. had the highest weight of butter in the Show.

In the open heifer milk test the first and second prizes went to Mr. Evens for Ruby XII. and Plenty V.

The following year, 1909, found Milker II. winning a second prize for Mr. Evens at the Oxfordshire Milking Trials, while at the Bath and West, at Exeter, Burton Fuchsia not only won in the milk test, but was first in the three days' butter test ; she gave 212lbs. 10oz. milk in the three days. At the Royal Counties Show, at Reading, Mr. Evens won first prize in the milk test, and second in the butter test with Tozzle, while Creamy IV. was second in the milk test and fourth in the butter test. First, second and third prizes in the breed milk tests at the Royal Show at Gloucester went to the Burton herd with Fuchsia III., Ruby IV. and Cork VI., while in the milk test open to all breeds and crosses, Tozzle was second and Fuchsia III. third, the latter also being third in the open special butter test. This cow also took a second prize at the Tring Milking Trials, while next in order to her came Ruby IV. and Bramblefinch from the same herd. Many honours fell to Mr. Evens and the Burton herd at the London Dairy Show that year, for Nancy V. stood first and Spotted V. second in the breed milk test, while in the open heifer milk test Pride V.

took a second prize. But Nancy V. was the cow to most distinguish herself, for besides winning in the Shorthorn butter test, she carried off the Lord Mayor's Cup, and the £50 Barkam Challenge Cup for the best milker in the hall.

This year, 1910, Mr. Evens again began his triumphs in dairy trials with a first at the Oxfordshire with Milker II., who gave 7½ gallons in 23 hours, while at the Bath and West, at Rochester, Amy was third in the milk test. In the breed milk test at the Royal Show, at Liverpool, first and second prizes went (for the sixth year in succession) to Burton cows, Fuchsia III. and Burton Cork VI., and they also stood first and second in the butter test open to all cows in the yard. Third in the breed tests came Mr. C. E. Scorer's Panton 206th, with a yield of 58lbs. 8oz., being 53 days in milk. She had been H.C. in the open butter trials. Mr. A. P. Brandt's Stenigot Bloom 10th, from the Bletchingley Castle herd, was reserve in the milk trials. It must be mentioned that Fuchsia III.'s yield of butter and milk constitutes a record for the Royal Show, and are really wonderful figures, for in 24 hours she gave 77lbs. 12oz. of milk, containing 4·5 per cent. of fat, which churned into 3lbs. 12½oz. of butter. A third in the milk test and a reserve ribbon in the butter test fell to Fuchsia III., at the Tring Dairy Trials; at the London Dairy Show she stood first and Ruby XII. third in the breed milk test, being divided by Mr. C. E. Scorer's Bracebridge No. 3,B., with Mr. G. L. Palmer's Bracebridge 79, B. reserve; while a fourth prize went to Fuchsia III. in the open butter test. In the milk test for Lincoln Red heifers Rose XI. took a first prize to Burton.

The Milking Records of the Bracebridge (Mr. C. E. Scorer's) and Burton (Mr. John Evens') herds are here appended :—

THE BRACEBRIDGE HEATH MILKING RECORDS
for 1909-1910.

Name of Cow.			Days in Milk.		Yield in Gls.
Bracebridge No. 102		..	456	..	1876
† ,, ,, 101B		..	245	..	473
,, Milkmaid		..	336	..	969
† ,, No. 94B		..	294	..	861
† ,, ,, 76B		..	259	..	601
† ,, ,, 0B		..	287	..	616
† ,, ,, 88B		..	287	..	632
,, ,, 42		..	301	..	1169
,, ,, 189		..	196	..	596
,, ,, 51B		..	231	..	858
,, ,, 199		..	198	..	630
† ,, ,, 194B		..	196	..	492
,, ,, 4B		..	154	..	459
,, ,, 40B		..	147	..	467
,, ,, 192		..	147	..	502
,, ,, 77B		..	161	..	471
Panton ,, 206		..	329	..	1206
Bracebridge ,, 118		..	301	..	1127
,, ,, 3B		..	301	..	1346
Cadeby Belle		..	266	..	1082
Bracebridge No. 17B		..	245	..	828
,, ,, 26B		..	238	..	893
† ,, ,, 121B		..	301	..	741
,, Fulletby		..	210	..	804
,, ,, 78B		..	238	..	912

25 cows and heifers milked averaged 824 gallons each.

The mark † denotes heifer's first calf.

BURTON MILK RECORDS FOR 1909-1991.

Cow's number in stall	Name of Cow	Date of Calving, 1909	Calf	Total yield in lbs.	Days in Milk	Average per day
20	Chance VII. ..	Jan. 5	1st	8,686	308	28·2
a11	Ruby Spot	,, 7	9th	9,739	294	33·2
b12	Blossom	,, 8	1st	10,801	385	28·1
18	Cowslip Red IV. .	,, 15	4th	7,909	280	28·2
15	Fuchsia II. ..	,, 29	5th	8,634	252	34·3
14	Buttercup II. ..	Feb. 4	1st	6,065	259	23·5
2	Fancy	,, 5	3rd	7,114	294	24·2
27	Marjorie II. ..	,, 9	3rd	9,582	294	32·6
c24	Amy	,, 14	4th	10,644	329	32·4
28	Quantity V. ..	,, 28	1st	8,163	329	24·9
38	Plenty III. ..	Mar. 8	4th	6,242	294	21·2
8	Susan	,, 9	2nd	6,285	266	23·6
19	Pride VII.	,, 11	3rd	8,661	266	32·6
d22	Vic IV.	,, 15	1st	8,163	329	24·8
e5	Milker II.	,, 18	2nd	10,153	259	39·2
10	Tulip IV.	,, 18	1st	8,989	343	26·2
3	Cowslip V.	Apr. 1	1st	7,933	301	26·4
f23	Cork VI.	,, 7	2nd	9,521	287	33·2
g41	Fuchsia III. ..	,, 13	3rd	11,863	308	38·5
h37	Primrose V. ..	,, 14	1st	9,234	350	26·4
i35	Princess	,, 17	2nd	7,135	287	24·8
40	Winter II.	,, 20	3rd	9,192	441	20·8
43	Bounty IV... ..	,, 27	6th	6,396	231	27·7
j39	Creamy IV. ..	May 3	4th	7,811	238	32·9
18	Violet VII. ..	,, 5	1st	4,937	224	22·0
44	Wanton II. ..	,, 11	3rd	7,467	259	28·9
k4	Tozzle	,, 22	5th	10,152	294	34·5
l7	Ruby IV.	,, 23	8th	12,213	322	37·9
I	Royal Maid II. ..	June 2	4th	7,190	238	30·4
m38	Cork V.	,, 8	3rd	8,059	294	27·5
30	Lassie	,, 11	3rd	8,026	294	27·3
n26	Ruby Spot III. ..	,, 13	4th	6,803	224	30·4
o11	Beauty II.	,, 21	4th	10,521	350	30·1
p8	Bramblefinch ..	July 16	4th	10,450	273	38·3
q6	Vic II.	,, 29	6th	9,139	301	30·4
40	Quality IV... ..	Aug. 2	2nd	10,462	455	23·0
r13	Ruby Spot II. ..	,, 3	6th	6,874	266	25·9
s2	Potentilla III. ..	,, 19	4th	5,599	210	26·7
t31	Nancy IV. ..	,, 23	6th	9,939	350	28·4
u15	Quality V.	,, 25	2nd	5,484	189	29·0
v16	Ruby XII.	Sept. 2	2nd	6,723	259	26·0
w4	Profit III.	,, 4	3rd	5,366	259	20·7
x9	Nancy V.	,, 7	5th	9,141	273	33·4
y36	Spotted V. ..	,, 13	4th	9,460	350	27·1
z14	Pride V.	,, 13	1st	11,664	315	37·0
28	Pride VIII. ..	,, 14	1st	7,680	329	23·4
ab39	Ruddy V.	,, 17	5th	7,655	266	28·8
34	Fox III.	Nov. 9	4th	9,438	322	29·3
5	Hyacinth	Dec. 1	3rd	6,681	259	25·8
19	Snowdrop	,, 2	1st	8,186	287	28·5
32	Aconite	,, 3	3rd	7,402	252	29·4

Fifty-one Cows yielded 427,626 lbs. Milk.
Average per Cow, 838·5 gallons.
23·5 per cent. were first calf Heifers.

31 Cows calving in 1890 averaged 740 gallons per Cow.

35	,,	,,	1891	,,	720	,,
34	,,	,,	1892	,,	795	,,
38	,,	,,	1893	,,	732	,,
39	,,	,,	1894	,,	834	,,
43	,,	,,	1895	,,	867	,,
43	,,	,,	1896	,,	889	,,
36	,,	,,	1897	,,	881	,,
38	,,	,,	1898	,,	824	,,
34	,,	,,	1899	,,	860	,,
36	,,	,,	1900	,,	785	,,
48	,,	,,	1901	,,	758	,,
40	,,	,,	1902	,,	776	,,
42	,,	,,	1903	,,	780	,,
43	,,	,,	1904	,,	842	,,
54	,,	,,	1905	,,	816	,,
48	,,	,,	1906	,,	802	,,
53	,,	,,	1907	,,	$771\frac{1}{2}$	,,
50	,,	,,	1908	,,	810	,,
51	,,	,,	1909	,,	$838\frac{1}{2}$	,,

PUBLIC AUCTION SALES.

PUBLIC AUCTION SALES.

Up to quite recently the Lincoln Reds have only been in the hands of Lincolnshire tenant-farmers till some ten years ago, when their merits began to force themselves on the notice of the shrewd business men in other countries, who desired a hardy, thrifty breed of cattle, combining great beef-producing with heavy milking qualities. No rich people have taken them up as a hobby, and Argentina is still so wedded to paper pedigree, that the purchases from that country have been comparatively few. Although the Lincoln Reds have been a registered and carefully-guarded breed since 1895, and they have been recognised as such, and classes granted them by the Royal Agricultural Society of England since 1901, the big estancieros will not, as a whole, look at a meritorious Shorthorn unless it has a lengthy C.H.B. pedigree. But this will not last much longer ; Lincoln Reds are going out to South America in greater numbers every year ; and when once the absurd prejudice against these sterling cattle is removed, prices will go up with a bound. A careful study of the annual tale told in the sale rings of Lincolnshire will convince the reader that prices are steadily rising year by year ; and as the cattle continue their uniform improvement in appearance, while maintaining and developing their great characteristics of early maturity, lean flesh, no waste, almost unrivalled milking powers, wonderful constitutions, and economy in feeding, so will they command the admiration they deserve and their value will become enhanced. The wise man, then, is he who will take time by the forelock and establish a registered herd while he can still do so at tenant farmers' prices, and not wait till he has to pay for a lengthy pedigree. Doubly wise, too, is he, who, if in touch with the milk demands of a big town, establishes a herd of Lincoln Reds that will not only produce a greater weight of milk of the required standard than any other breed in existence, but can command good prices in the sale ring and the cattle market. I have referred to what Mr. John Evens' herd has done against all breeds in milking trials and butter tests ; and his successes at the Lincoln Fat Stock Shows, and in the show-yards of all England, demonstrate that beef can be produced even in a famous dairy herd, when that herd is one of the Lincolnshire Red breed. Can there possibly be a better dual-purpose cattle than these, then ; can the tenant farmers of England wish for a better friend ?   I have already quoted the old saying of " Booth for the butcher and Bates for the pail."   The time is not far distant when it will be generally recognised that it is a case of " the Lincoln Red for the butcher, and the Lincoln Red for the pail ; the Lincoln Red for the producer and the Lincoln Red for the consumer."

1896.   I will begin my story of the sale ring with the time when the breed first became registered, and the Lincolnshire Red Shorthorn Association first took the annual Bull Sales, at Lincoln April Fair, under their management.   That the move was a good one, and that it has proved a boon and a blessing to the farmers of Lincolnshire, is amply demonstrated by the improved prices, the increased entry of bulls at the sales, and the additional buyers from more distant countries who come year by year.   The first annual sale of the Association was conducted at Lincoln, on April 23rd, *1896*, and it will be noticed how, without any sensational prices, a uniformly satisfactory and gradually-increasing average has been maintained from year to year.   At this initial sale of the Association then 140 bulls changed hands for a total sum of £3,285 19s. 6d. and at an average of £23 9s. 5d., which was distinctly satisfactory when it is considered that the highest individual price was but 50gs. Mr. George Walker, Bigby, sold two sons of Bigby (319) at an average of 35gs., one, Bendigo (312) going to Mr. W. Martin, Wainfleet, at 40gs. ; and Messrs. W. T. Wells & Sons sold 11 at an average of £22 11s. 6d., Mr. Robert Walker, Aylesby Manor, giving 34gs., for the pick of the bunch Hag Jumbo (394) ; one of the bulls offered by Messrs. J. H. & R. Wright, Hagworthingham, went to Mr. L. W. Stephenson, South Thoresby, at 40gs. ; and the four bulls sent by Mr. G. Marris, Kirmington House, Brocklesby, averaged £39 5s., Mr. H. B. Minta, Normanton, and Mr. J. Ward, Moulton, being purchasers at 43gs. and 40gs. respectively ; while Mr. Pereira Brown, Glentworth Hall, sold Glentworth Rufus 6th (389) to Mr. J. Cartwright, Dunston Pillar, at 43gs. and Glentworth Rufus 7th to Mr. C. Minta, at 50gs.   Both these bulls were by Glentworth Rufus (128) a son of Eclipse (111) who was bred by Mr. W. Chatterton, Hallington.   Mr. G. E. Sandars, Fillingham Manor, sent up five, whose average was £27 18s. 7d., the pick being Red Knight (924), by King Hal (156), knocked down to Mr. W. W. West, Needham Hall, Wisbech, at 30gs. ; and the ten from the Dowsby Hall herd belonging to Messrs. S. E. Dean & Sons, averaged £30 6s. 10d., the highest price being 44gs. which Mr. E. Mowbray, Fishtoft, paid for Dowsby Red King (365), a bull descended from the strains of Mr. J. Parker, Ingleby. Mr. John Ruston, Hemswell, sent a quartet which realized an average of just 31gs., one of them Hemswell Baron (404), by Laughton Prince XXVII (581), a bull of C.H.B. parents, going to Mr. Tong, Stragglethorpe, at 40gs. The average for the eight belonging to Mr. Reuben Roberts, Horncastle, was £26 6s. 3d., his highest price being 38gs. ; as was that obtained for the best of the four sent by Mr. J. W. Measures, Dunsby, Bourne. Mr. E. H. Cartwright's five from Keddington, averaged £33 9s. 10d., most money being given for Rear-Admiral (920), by Admiral (3), who was knocked down to Mr. Pereira Brown, Glentworth Hall, at 45gs. Another bull was knocked down at 50gs., this being Saltfleet Saturn (503), by Saltfleet Eclipse (227), the property of Mr. T. B. Freshney, South Somercotes, who was purchased by Messrs. R. & W. Wright, Nocton.

1897. At the second sale held by the Association, in April Fair week, *1897*, no fewer than 225 bulls were sold, their average price being £21 0s. 1d. and the total aggregate £4,521 16s. 6d. The supply rather exceeded the demand, and as there was no upset price, several bulls changed hands at single figures. But there was more money for the best bulls, and the sale was distinctly encouraging. Messrs. W. T. Wells & Sons, Withern, sold seven at an average of exactly £30, Mr. G. A. Brown, Maidenwell, paying 44gs. as top price ; whilst the half-dozen sent up by Mr. E. H. Cartwright, Keddington, obtained an average of £32 4s., most money being given for Keddington Saturn (803), by Astronomer (21), who was knocked down to Mr. J. W. Rowland, Fishtoft, at 46gs. Messrs. R. & R. Chatterton's quartet from Stenigot averaged £41 4s. 3d., Messrs. Tomlinson & Hayward giving 48gs. for Meadow King (863), by Commander (80). The highest price of the day was 60gs., which Mr. J. W. Measures, Dunsby. paid for Saltfleet Sentinel (945), one of the five sent up by Mr. T. B. Freshney, South Somercotes, whose average was £37, and another from the same herd, Saltfleet Rodney (943) made 56gs., to Mr. Riggall, Well. Bullbreeder's Eclipse (647), by Conisholme (82), the solitary representative from the herd of Mr. John Searby, Croft, went to Mr. E. Turnor, Panton, at 50gs. ; Conisholme Samson, the property of Mr. J. W. Hill, Smethwick Hall, Congleton, was knocked down to Mr. J. W. Ward, Withcall, at 57gs. ; and Mercury (1443), by Marmion (C.H.B.

65,892), belonging to Mr. W. J. Atkinson. fell to Mr. S. E. Dean, Threekingham, at 48gs., while Lincoln Waterloo 9th (718), by Dowsby Waterloo Duke (102), one of the string sent up by Messrs. S. E. Dean & Sons, Dowsby Hall, fell at 52gs. to the bid of Mr. J. Langham, Hough Manor, Mr. T. Atkinson, North Kelsey, paying 45gs. for Cambridge Duke 19th (707) from the same herd. The ten belonging to Mr. Reuben Roberts, Horn-castle, had an average of £27 18s. 7d., Mr. W. Barnes, Ewerby, giving most money, 47gs., for Digby Rose (705), by Stenigot Rose (533).

1898. Next year, *1898*, the upset price was fixed at 15gs., and the average in consequence rose to £22 5s. 3d. for 193 lots, though a little over £1 below that of the first sale. Messrs. W. &. H. W. Scorer sent up five from Burwell, Louth, the average being £32 11s. and the highest-priced animal Burwell Benedict (1110), sold to Mr. Colin MacIver, Ludford, at 43gs.; but most money on the day was found for Revolving Light (1658), a bull who afterwards proved of the greatest service in the herd of his purchasers, Messrs. T. & W. Dickinson, Worlaby. He was the property of Messrs. R. & R. Chatterton, and was bred at Stenigot, being by Coastguard (343) from a Commodore (87) cow, and he was indeed a great bargain at 62gs. Another of Messrs. Chatterton's bulls, Lord Salisbury, was sold to Mr. W. Bowser, at 45gs., and the eight lots averaged £37 17s. 3d. Mr. T. B. Freshney's five bulls averaged £34 19s. 3d., Saltfleet Toro (1343) being knocked down to Mr. S. E. Dean, Three-kingham, at 45gs.; and another famous bull sold at this sale was Conisholme Boy (347), the property of Mr. J. B. Hill, Smethwick Hall, Congleton, who was purchased by Messrs. R. & R. Chatterton, at 54gs. This afterwards great sire was bred by Mr. R. N. Sutton-Nelthorpe, at Scawby Hall, in 1893, and was by Glengarry (C.H.B. 62,653), from a cow bred by the late Mr. James Martin, Wainfleet.

1899. In *1899* fewer bulls were sold, but much better prices were obtained, there being a keen demand, while the quality of the animals was excellent, 147 lots changing hands at an average of one penny short of £28. Mr. J. W. Measures sent up three from Dunsby Grange, the average for which was £41 6s., the top price being 45gs., which Mr. G. Marris, Kir-mington House, paid for Dunsby Swell (1537), by Saltfleet Sentinel (945); and Mr. T. Bett's Benniworth 7th (1458), by Asterby Red (19) was sold to Mr. C. Bowser, Holbeach Marsh, at 70gs. Seven bulls from the herd of Mr. Reuben Roberts, Horncastle, averaged just £36, the pick of the lot being Digby Cavalier (1510), by Sotby Red 9th (516), who fell to the bid of Mr. John Todd, Kirkby Green, at 73gs. A string of ten represented the famous Stenigot herd, and these obtained an average of £36 6s. 7d., most money being forthcoming for England's Beau (1882), a son of Wolseley (1436) and Stenigot Belle, who became the property of Messrs. Simons Bros., Sutton-on-Sea, at 63gs.; while four fine bulls belonging to Messrs. T. & J. B. Freshney, South Somercotes, averaged £56 3s. 6d., the top price of the day being given for Saltfleet Actor (1664), by Shooting Star (1674), out of a Saltfleet Eclipse (227) dam, who became the property of Mr. T. Bett, Benniworth, at 81gs.; and Saltfleet Admiral (1665), by Saltfleet Mars (1342) fell at 60gs. to Mr. W. R. Caudwell, Holbeach. The best of the five sent by Messrs. S. & J. W. T. Crawley, Hemington, Oundle, was Royal George (1661), by Baron Ormsby 3rd (26), who went to Mr. Ireland, Marshchapel, at 50gs.; and Lord Glengarry (1926), by Keelby Dab (804), bred and owned by Mr. J. E. Davy, Tathwell, went at 58gs. to Mr. W. Ward, Leadenhall. Messrs. S. E. Dean & Sons, Dowsby Hall, obtained an average of £34 3s. 8d. for their nine bulls, the highest price being 65gs. which Mr. L. W. Stephenson, South Thoresby, paid for Dowsby Virtuoso 15th (1525), by Virtuoso (C.H.B. 69,763), out of a cow by Cambridge Duke 30th (C.H.B. 60,491).

1900. The next year, *1900*, in spite of a record for individual price thus far in the history of the sales, the average dropped to £23 18s. 6d.; but there were 49 more bulls sold than in the previous year, the actual number being 196. Mr. T. Bett's pair averaged £47 15s. 6d., Benniworth 10th (1774), by Fire Brick (1199) being knocked down to Mr. Reuben Roberts at 51gs., while Benniworth 8th fell to the bid of Mr. R. Mowbray, Gosberton, at 40gs. Mr. Scott, Clayworth, gave 50gs. for one of the eight sent by Messrs. W. T. Wells & Sons, Withern, and Digby Sotby 2nd (1845) and Poolham Butterman 5th (1974) were respectively knocked down to Mr. G. E. Sandars, Scampton, and Mr. A. Mountain, South Owersby, at 35gs. Mr. G. E. Sandars' seven averaged £35 5s., Mr. Reuben Roberts giving most money for Scampton Abbot (2325), by the Keddington-bred Great Tom of Lincoln (392) from a King Hal (156) dam, of the best Hallington blood, who was knocked down at 45gs.; while Mr. W. Sargeant, Ranby, paid 40gs. for

F

Scampton Admiral (2326) who was similarly bred. Great Tom of Lincoln himself, then five years of age, was sold to Messrs. Tomlinson & Dean, at 30gs. For Kirkby Middle Marsh II. (1918), by Middle Marsh (182), the property of Mr. John Todd, Mr. Vergette of Peterborough paid 50gs.; and Mr. W. J. Atkinson, Weston St. Mary, paid 46gs. for Messrs. S. E. Dean & Sons' Dowsby Virtuoso 26th, by Virtuoso (C.H.B. 69,763). Messrs. T. & W. Dickinson, Worlaby, obtained 45gs. for one of their bulls, Mr. J. E. Davy, Tathwell, being the purchaser ; but the highest price of the sale was obtained by Messrs. S. & J. W. T. Crawley, Hemington, Oundle, whose fine bull, Bumper 2nd (1703) was sold to Mr. W. B. Swallow, Wootton Lawn, Ulceby, at 105gs. He was bred at Hemington, being by Baron Ormsby 3rd (26), who was bred by Messrs. Mundy & Ward and was descended frow the famous herd of Mr. Oliver, Eresby. Saturn (2010) by the same sire, was sold to Mr. P. F. Brown, Glentworth Hall, at 50gs., and the five from Hemington averaged a shade under £47. Another promising young bull was Somercotes Star (2013), by Shooting Star (1674), the property of Mr. T. Wells, South Somercotes, who became the purchase of Mr. H. Young, Beesby, at 52gs.

1901. The bulls were sold at an average of £25 10s. 11d. in *1901*, 164 changing hands. Mr. W. B. Swallow, Wootton Lawn and Horkstow, sold four at an average of £30 9s., and Mr. T. Bett, five at an average of £34 0s. 5d. ; while Mr. Reuben Roberts disposed of 10, the average for which was £33 19s. 4d., the top figure of the day being obtained for Digby Conqueror (2182), by Calceby Stencil (1117) of Mr. John Mason's strains, from a cow by Stenigot Rose (533) of Messrs. Chatterton's breeding, who was bought by Mr. G. E. Sandars, Scampton, at 85gs. Mr. Sandars himself had the satisfactory average of £37 4s. 4d. for his nine, Mr. L. W. Stephenson, South Thoresby, and Mr. H. R. Sharpe, Swineshead, respectively, giving 46gs. each for Scampton Bloodsucker (3043) and Bloodstone (2633), both out of King Hal (156) dams ; while the top price of the bunch was obtained for Bloodstain (2632) by the same sire as the other two, but out of a cow by Calceby (51), who became the purchase of Mr. J. G. Williams, Pendley Manor, Tring, at 60gs. Lord Heneage gave 50gs. for Mr. E. H. Cartwright's Keddington Kinsman (2243), by Bigby (319) out of an Admiral (3) cow ; and Messrs. R. & R. Chatterton disposed of three bulls at an average of £56 14s., one of them being the afterwards successful sire Lord Wolseley

(2572), by Wolseley (1436), from Stenigot Cowslip, by Comet (79), who went into the well-known herd of Messrs. T. & W. Dickinson, Worlaby, at 72gs., while Mr. W. S. Fox, Potterhanworth, paid 50gs. for Bloemfontein (2115), by Wrangler (C.H.B. 71,901). Mr. T. B. Freshney also had the good average of £38 13s. for his five bulls, most money being made of Saltfleet Conqueror (23,211), by Grandad (1561) who was bought by Mr. R. S. Hand, Croft at 68gs., and the next best price being given by Mr. W. Goddard, Nottingham, for Saltfleet Colonel (2318), by Saltfleet Bravo (2006), who realized just 50gs. A smart bull bred by Mr. John Abraham, Walesby House, was also knocked down to Messrs. R. & R. Chatterton at 52gs.

An Autumn sale under the auspices of the Association was held at Lincoln that year, but was not attended with any great success, and after one more attempt it was dropped in 1903. The first of these two sales was brought off on October 18th, 1901, the highest price for cows and calves being 29gs., which Mr. W. T. Bagshaw, Bolsover, paid for a cow belonging to Messrs. T. & W. Dickinson, Worlaby ; while most money for bulls was forthcoming for Mr. J. Evens' Royal Burton (1996), a famous prize-winner, who went at the ridiculously low price of 40gs. to Mr. C. Tinley, South Hykeham.

The first of the newly-organized sales at Alford November Fair resulted that year in 56 bulls being sold at an average of 15¼gs.

1902. In *1902* an average of £25 4s. 4d. was obtained for 169 bulls at the seventh Spring sale, held at Lincoln, on April 24th, and this in spite of the fact that the highest individual price was but 60gs. Mr. J. Measures, Dunsby, sold the best of his lot, Candidate, to Mr. J. Evens, Burton, at 40gs., and Brilliant Star (2428), the property of Messrs. R. & R. Chatterton, was knocked down to Mr. Coleman, Bingley Grange, at 47gs. Messrs. T. & J. B. Freshney had an average of £28 4s. for the eight bulls from Somercotes, and the Exors. of the late Mr. John Abraham, Walesby House, sent up three, the average for which was £35 14s. Lord Bigby, bred by Mr. J. E. Davy, Tathwell, made 42gs. to Mr. J. Sharpe, Bardney, and among the purchases by Mr. J. S. Gordon, for the Congested Districts Board, Ireland, were two bulls belonging to Messrs. T. & W. Dickinson, at 47gs. and 38gs., the average for the four Worlaby bulls being £37 10s. 9d. Mr. T. L. Thurlby, Cranwell Lodge, Sleaford, gave 45gs. for Welbourne

Baronet (2700), the property of Mr. J. C. Mountain, Welbourne, and Mr. J. W. Farrow's Strubby Red Coat B (2659) was sold to Mr G. W. Harvey, Londonthorpe, at 60gs., the same price being made of Mr. G. E. Sandars' Scampton Cyclone, who went to Mr. R. Mowbray, Gosberton. Mr. Farrow's bull was by Red Curly Coat (923) from a Thumper (550) dam, while Scampton Cyclone (3045), was by Great Tom of Lincoln (392), out of a cow by King Hal (156). The five Scampton bulls averaged £41 7s. 4d., and four sent by Mr. T. Bett, Benniworth, £45 8s. 3d., the best average of the sale. Mr. J. S. Gordon took Benniworth 17th (2410) and Benniworth 19th (2412) back to Ireland with him, and Benniworth 19th (2411) made 50gs. to Mr. Humphreys.

The second Autumn sale of the Association, held that year on October 17th, met with no better success than the one held the previous year, and it was realized that it was fulfilling no useful purpose. Mr. George Cartwright, Ragnall Hall, Newark, sold Branston Duchess 2nd to Mr. Ellerby, Fledborough, at 26gs. ; and the highest price bull was Mr. J. C. Mountain's Welbourne Ormsby (3139), bought by Mr. J. Evens at 35gs.

The number of bulls sold at Alford in November this year was 56, and they averaged 20gs.

1903. The eighth Spring sale of the Lincolnshire Red Shorthorn Association was held at Lincoln, on April 23rd, *1903*, 189 bulls changing hands at an average of £26 5s. 3d., a great improvement on the previous year, when it is remembered that 20 more bulls were sold. Mr. B. Simons, The Grange, Willoughby, sold three at an average of £34 6s., Walmsgate Boatswain 7th (3129) going to Mr. W. Hoff, Grebby, at 44gs. ; while Mr. Reuben Roberts sold 8 at an average of £44 10s. 10d. Digby Scampton 1st (2864) and Digby Scampton 2nd (2865), both by Scampton Abbot (2325), from cows by Sotby Red 9th (516), each making 52gs., Mr. E. H. Cartwright buying the first, and Mr. F. Higgins, Alford, the other. Mr. G. E. Sandars' half-dozen from Scampton averaged £32 7s. 6d., most money being made of Scampton Drama (3958), by Digby Conqueror (2182), who was knocked down to Mr. J. Marriott, Cropwell Butler, at 45gs. Messrs. T. & W. Dickinson's ten averaged £29 1s. 8d. and Messrs. T. & J. B. Freshney's six £31 10s., Saltfleet Echo (3038), by Grandad (1561), from a cow by Saltfleet Sappy (502) making top price of the day when sold to Mr. T. Bett, Benniworth, at 65gs. The Exors. of the late Mr. J. Abraham obtained an average of £49 12s. 3d. for their four

capital bulls, one of which Otby Moonshine (3211), by Otby
Eclipse (2286) went to Mr. J. W. Rowland, Fishtoft, at 54gs.,
while Walesby Woldsman (3684) and Walesby Watchman
(3683) were sold at 50gs. each respectively to Mr. J. Raynor,
Mansfield, and Mr. Topham, Thorney Park, Peterborough,
Half-a-dozen Hallington bulls belonging to Mr. W. Chatterton
averaged £32 4s. ; and the same number offered by Mr. W. B.
Swallow, Wootton Lawn, Ulceby, averaged £30 12s. 6d.,
Captain the Hon. G. B. Portman, Healing Manor, giving 40gs.
for Horkstow Diamond (2919), by Bumper 2nd (1793).

That year sales were started at Louth, and the Alford sales
were continued, those at the former place being held on October
22nd, when satisfactory prices were obtained. A cow belonging to Mr. W. Chatterton, Hallington, went to Mr W. J.
Atkinson, Weston St. Mary, at 56gs. ; but perhaps the most
interesting sale of the day was the disposal of Mr. T. B. Freshney's Saltfleet Favourite to Mr. W. J. Atkinson, at 40gs. This
heifer, who was calved in March, 1901, was by Grandad (1561),
out of a cow by Monarch (1292), whose dam was the famous
Nonpareil, by Hyllus (149), her dam being by Red Prince (211)-
and grand-dam by Crœsus (155). Saltfleet Favourite after,
wards won innumerable prizes both for Mr. Atkinson and Mr.
John Evens at the Royal, Bath and West, Peterborough,
Lincolnshire and other shows. Healing Miss Patrick, also from
Mr. Freshney's herd, made 33gs. to Mr. H. Abraham, Walesby,
and her bull calf went to Mr. J. Mason, Calceby, at 36gs. ; Mr.
H. Abraham also paying 30gs. for Healing Miss Wainfleet,
while Mr. J. Marriott bought her heifer calf at 25gs. Four
heifers and their calves, and Saltfleet Favourite, together
averaged £47 15s. 6d.; and four heifers belonging to Mr. G. A.
Oliver, Hallington, £24 16s. Mr. J. Marriott sent a long string
from Nottinghamshire, getting 28gs. for the heifer Needham
Wild Eyes, and 28gs. for Birthday (1782).

Good prices were also obtained at Alford on November
8th, where Messrs. J. W. Farrow & Sons, Strubby Manor,
obtained an average of £27 7s. for 12 young bulls, the highest
price of the day being 61gs. given by Mr. Clayton, Littleworth,
while Mr. C. O. Parr, Well, and Mr. S. E. Dean, Threekingham, paid 45gs. and 40gs. respectively for Strubby bulls.
Croft Sirdar (2842), the property of Mr. J. Searby, was also
sold to Mr. Scambler, Conington, at 42gs. The average price
for 95 bulls was 16½gs.

1904. In *1904*, the ninth sale of the Association was held at Lincoln, on April 28th, when 219 bulls were sold at an average of £25 4s. 5d., the highest price being 130gs., while another was sold for 100gs. Mr. W. B. Swallow sent up eight, the average for which was £28 12s. 3d., Mr. P. F. Brown, Glentworth Hall, paying 46gs. for Horkstow Herald (3919), while Mr. W. Chatterton, Hallington, gave 41gs. for Horkstow Hero (3422), both being by Bumper 2nd (1793) for whom Mr. Swallow paid Messrs. S. & J. W. T. Crawley, Hemington, Oundle, 105gs. at the 1900 sale. Mr. Tweed, of Lincoln, gave 44gs. for Horkstownian Alford 2nd (3427), the property of Mr. E. J. Turton, Horkstow ; and Mr. T. B. Freshney's Barnum (3576) was knocked down to Messrs. W. J. Coleman & Sons, London, at 47gs. Otby Reduction (3513), belonging to Mr. E. Arbaham, Otby, made 56gs. to Mr. W. S. Fox, Potterhanworth, and Otby Reformer (3514) fell at 50gs. to the bid of Mr. F. Riggall, Croxton. Both these bulls were by Otby Eclipse (2286). Lord Heneage's Hainton Gunner (3401), by Keddington Kinsman (2243) also made 50gs., being purchased by Mr. J. H. Gaunt, Waddingworth ; while Mr. Tom Bett's Benniworth 28th (3214), a son of that fine bull Red Chief (2611) and a Benniworth Kelstern (28) cow, was knocked down at 55gs. to Mr. F. Higgins, Alford. One of Mr. Reuben Roberts' bulls, Digby Scampton 8th (3353) was sold at 42gs. to Mr. W. R. Caudwell, Holbeach Marsh, and Mr. J. W. Measures' Bonny Boy (3758), by Western Nonpareil King (2068) became the purchase of Mr. Reuben Roberts at 58gs. But both the highest individual price and the best average went to the Scampton herd, as Mr. G. E. Sandars' ten bulls averaged £61 8s. 6d., and one of them was knocked down to Mr. John Evens, Burton, at 130gs. This was Scampton Expansion (4093), a winner afterwards both at the Royal and Lincolnshire Shows, and a most successful sire. He was (as indeed were all Mr. Sandars' bulls) by Keddington Ruby (1243), who was by Bigby (119) first both at the Royal and Lincolnshire Shows in 1898, to whom Keddington Ruby was second on each occasion, and he was out of a cow by Great Tom of Lincoln (392), whose dam was by Hallington (135), and grand-dam by Cawkwell (67). Messrs. T. & W. Dickinson and Mr. J. Mason both bid well for this bull. A similarly bred bull from the same herd, Scampton Executioner (4090) made 100gs., going to Mr. J. Mason after a duel with Mr. Reuben Roberts ; and good prices were also obtained for Excursionist (4089) who went to Messrs. J. & W. Dickinson, Worlaby, at 62gs., and Explorer (4095) who became the property of Messrs. T. Atkinson & Son, North Kelsey, at 52gs. Two other Scampton bulls were knocked down at over 40gs. each. Mr. Gordon again made numerous purchases for the Irish Congested Districts' Board, and the Duke of Devonshire was also a good customer.

Another sale was held at Louth that year, on October 20th, prices again ruling satisfactory. Messrs. R. & R. Chatterton's Stenigot Dairymaid went to Mr. B. G. Stone, Elkington, at 40gs., and Mr. T. B. Freshney's Ruby 14th, a member of a most famous family and by Regent (C.H.B. 73,398), was sold to Mr. W. J. Atkinson, at 50gs. Mr. E. H. Cartwright's bull calf by Vanguard (2691), out of Wells No. 5, went to Mr. W. Chatterton, at 60gs. ; while Vanguard himself, who scaled at the time 1 ton 7 cwt., was bought by the Messrs. Chatterton for the Argentina at 40gs.

The fourth annual sale at Alford was brought off on November 8th, and proved highly successful, 120 bulls changing hands at an average of 19gs. Mr. John Mason's Calceby Master (3275) went at 55gs. to Mr. J. Bowser, Frithville, and Mr. G. E. Sandars gave 48gs. for Saleby Premier (4079) belonging to Messrs. S. & J. H. Briggs, Saleby. But the highest price up to that time given for a Lincoln Red bull at a public auction was obtained for Mr. Tom Bett's Benniworth 29th (3215), who was sold to Messrs. R. &. R. Chatterton, at 150gs. He was by Red Chief (2611) from a cow by Benniworth Kelstern (28), was champion at the Alford Show that year, and was respectively second and third at the Royal and Lincolnshire Shows. Mr. J. Searby's Croft Marvel (3829) made 42gs. to Mr. J. Byron, Normanby ; and Mr. C. W. Tindall's Pure Gold, by Guinea Gold 2nd (3393) went to Mr. Nainby-Manby at 50gs. ; while Messrs. R. & R. Chatterton's Stenigot Beau (2641), by Wolseley (1436) was sold to Mr. T. Bett, at 60gs. Tothby Virtuoso 2nd (4167), belonging to Mr. G. J. Brown, Tothby, was sold to Lord Willoughby de Eresby, at 52gs.

1905. The next year, *1905*, the tenth sale of the Association was held at Lincoln, on April 27th, when 187 bulls were disposed of, the average being £24 11s. 3d., and slightly lower than that of the two previous years. Mr. G. E. Sandars again obtained both the highest price and the best average with his Scampton bulls, one of which, Scampton Fortress (4563), by Keddington Ruby (1243) from a cow by Great Tom of Lincoln (392), was sold to Mr. L. W. Stephenson, South Thoresby, for 100gs. ; while For Ever (4558), similarly bred, was sold to Mr.

F. J. S. Brown, Sedgeford, at 86gs. ; Forager (4555) went to
Mr. H. Caudwell, Midville, at 52gs. ; and Mr. G. Marris, Kirmington, and Mr. J. G. Barlow, Burton, gave 42gs. and 41gs. .
respectively for Forester (4537) and Force. Mr. Sandars' ten
good bulls averaged £52 10s. Mr. R. C. Bemrose, Caythorpe,
gave 48gs. for Threekingham Gay Lad (4164), by Great Tom of
Lincoln (392), belonging to Mr. J. Todd, Kirkby Green ; and
Sharpshooter (4091), by Field Cornet (2595) was knocked down
to Mr. J. Cartwright, Dunston Pillar, at 76gs.   He belonged to
Mr. W. S. Fox, Potterhanworth, whose three bulls averaged
£52 17s. Mr. Reuben Roberts sold eight at an average of
£27 11s. 3d. ; Mr. T. Bett four at an average of £33 6s. 9d. ;
Mr. W. Drakes four at an average of £28 12s. 3d. ; Messrs. T.
& W. Dickinson ten at an average of £26 2s. 10d. ; Messrs.
R. & R. Chatterton five at an average of £28 15s. 4d. ; and
Mr. J. Measures five at an average of £28 7s.

There were no high prices at the third sale at Louth, on
October 5th, 1905, but Lord Egerton of Tatton purchased
several useful heifers, and a bull calf by Saltfleet Bonus (3582),
out of Keddington Lady Bird 6th, was bought by Mr. Harvey,
Londonthorpe, Oakham, at 41gs.

Better prices were forthcoming at the fifth Alford sales
held on November 7th.   The previous evening there had been
a big gathering of Lincoln Red breeders to meet at dinner
the Duke of Portland, the Right Hon. A. E. Fellowes, Minister
of Agriculture, Lord Willoughby de Eresby, M.P., and others.
The highest price at the sale was obtained by Messrs. R. & R.
Chatterton for Stenigot Primate, by Red Chief (2611), from
Red Daisy, by Stenigot Wrangler (2341), who was sold to
Mr. G. H. H. Kennedy, to go to Buenos Ayres, at 165gs.   He
had taken a first at Peterborough, and second at the Royal,
and a third prize at the Lincolnshire Show.   Messrs. Parker
& Frazer, Liverpool, also gave 46gs. for Stenigot Gwynne, by
Red Chief (2611).   Mr. T. Bett made 51gs. of Benniworth
38th (3753), by Keddington Kinsman (2243), to Mr. Harvey,
Londonthorpe, and Mr. F. Scorer, Bracebridge, secured Messrs.
T. & W. Dickinson's Bonby Kinsman (4251) also by Keddington Kinsman, at 50gs.   Thirty-six bulls made 20gs. and
upwards, and the average for 86 was 21½gs.

1906.   Three records were set up at the 11th sale of the Association, at Lincoln, on April 26th, *1906*, marking a distinct epoch in the history of the breed, as a bull was sold to go
to Chili, at 305gs. the highest price so far made of a Lincoln

Red in England ; 166 animals were sold at an average of £27 10s. 9d. and seven from one herd at an average of £89 5s. To Mr. G. E. Sandars, Scampton House, Lincoln, belongs the first and third of these records, and the bull in question was Scampton Goldreef (4569), who was purchased by Mr. James Cartwright, Dunston Pillar, acting on behalf of Mr. Cockbain. Mr. J. W. Dennis, Boston, tried to keep him in the country, but he had to give way before the weight of foreign money. Goldreef was a typical Lincoln Red, combining lean flesh with substance, quality and colour ; and his breeding was of the best, for he was by Keddington Ruby (1243) from a cow (a great milker), by Great Tom of Lincoln (392), whose dam was by King Hal (156), grand-dam by Hallington (135) and great-grand-dam by Cawkwell (67). Mr. Dennis had to be content with Gold-mine (4567) by the same sire, which cost him 90gs., and the Earl of Warwick, and Mr. Minton, Apethorpe, gave 56gs. and 50gs. respectively for Goldseeker (4570) and Lord of the Valley, the first-named, bred like the two big-priced bulls, and the second by Digby Captain (2505), of Messrs. C. & R. Brooks' breeding. Colonel C. A. Swan, Sausthorpe, sold five bulls at an average of £60 5s. 4d., Mr. J. W. Measures paying 95gs. and Mr. A. Turner, Oadby Pastures, Leicester, 56gs. respectively for Sausthorpe Red 12th (4552) and 13th (4553), both by Bonby Wolseley 17th (3233), while Mr. G. E. Sandars secured Sausthorpe Red 10th (4550) at 72gs. This latter bull after being used in the Scampton herd was sold to go to South Africa, where, at the Bloemfontein Show, in 1907, he won the championship of all breeds. Messrs. T. & W. Dickinson had an average of £32 15s. 2d., the highest price being 52gs. which Messrs. T. Atkinson & Son paid for Bonby Kinsman 9th (4700), by Keddington Kinsman (2243).

Prices ruled low at Louth, on October 28th ; but there was the good average of £25 18s. 7d. for 104 bulls sold at Alford, on November 7th, while 12 females averaged £21 4s. 4d. Mr. J. Langham's Royal winner Brandon Grenadier (4274) was bought by Mr. G. E. Sandars to go into the Scampton herd at 200gs., Mr. G. J. Brown, Tothby, and Mr. S. B. Carnley, both being keen opponents. He was a most symmetrical bull of true Shorthorn type, very even in flesh and a good colour, and he was by Brandon Chief Justice (2425), from Brandon Nonpareil, by Chancellor (332), out of Keddington Wainfleet, by Bigby (319), Keddington Wainfleet's dam being Wainfleet 2nd, by Solferino (C.H.B. 59,992). Mr. J. W. Dennis, Boston, paid 88gs. for Keddington Baron (4881), by

Saltfleet Bonus (3582), out of Keddington Baroness, by Benniworth 4th (629), and Keal Tommy (4880), by Scampton Excavator (4084), was sold to Mr. W. Dodds, Donington, at 68gs. Mr. J. Searby's Croft Viol (4332) made 55gs. to Mr. R. Mowbray's Exors., Gosberton, and Mr. G. J. Brown's Kirkby Virtuoso (2947) went at 50gs. to Mr. H. Young, Beesby. The young heifer Trusthorpe Fairy, by Buscot Loyalty (C.H.B. 80,591), belonging to Mr. B. Simons, was knocked down to Mr. S. B. Carnley at 40gs.

A bull sale was held at Boston, on May 4th, 1906, when 51 lots changed hands at an average of £20 8s., most money being made of the late Mr. T. Bett's Benniworth 39th (4239), by Saltfleet Echo (3038), who was sold to Mr. T. P. Horn, Heckington, at 70gs.

1907. The 12th sale of the Lincolnshire Red Shorthorn Association was held on April 25th, 1907, when an average of £26 2s. 11d. was obtained for 210 bulls. For the twelfth time Mr. G. E. Sandars obtained the highest average at the Lincoln bull sales, and for the tenth time he secured the highest individual price, the eight Scampton bulls averaging £57 1s. 10d., while one of them, Hermes (4972), the first prize-winner in the older bull class (this being the first year that prizes were offered by the Association), went to Mr. W. B. Swallow, Wootton Lawn, Ulceby, at 140gs. Scampton Hercules (4969) was also sold to Mr. A. Mountain, South Ormsby, at 54gs., and Herald (4968) to Mr. F. Scorer, Nettleham, at 50gs., both these bulls being by Keddington Ruby (1243), from Great Tom of Lincoln (392) dams, while Hermes was by Keddington Ruby, from a cow by Digby Conqueror (2182). Mr. Reuben Roberts sold eight at an average of £43 3s. 7d., Mr. W. R. Sharpe, Swineshead, paying 62gs. for Thimbleby Curly Coat, by Scampton Excelsior (4085); and Colonel C. A. Swan's four averaged £42 5s. 3d., most money being paid for Sausthorpe Red 15th (4961), by Bonby Wolseley 17th (3233), who became Mr. H. Caudwell's property at 58gs. The ten from the herd belonging to Messrs. T. & W. Dickinson averaged £34 19s. 3d., the biggest bid being forthcoming for Bonby Excursionist 4th (5161), by Scampton Excursionist (4089), who was purchased by Mr. J. G. Williams, Pendley Manor, Tring, at 65gs. The Exors. of the late Mr. T. B. Freshney obtained 62gs. for Saltfleet Dragoon (4547), by Saltfleet Bonus (3582), Mr. G. J. Brown, Tothby, being the purchaser; and Messrs. J. W. Farrow & Sons sold Strubby

Red Coat 1st (5647), by Under-Porter (3126) to Mr. H. Abraham, Walesby, at 52gs.

At Boston, on May 4th, 1907, 74 bulls were sold at an average of £21 4s. 6d., Mr. Reuben Roberts' Bucknall Captain (5199), by Scampton Excelsior (4065) going to Mr. F. Scorer, Bracebridge, at 42gs.; and Messrs. J. N. Robinson & Son disposing of Anderby Bumper, by Keddington Surprise (3445) at 41gs. to Mr. J. T. Smithson, Tattershall, Thorpe.

At Alford, at the seventh annual bull sale, held on November 7th, 1907, the average for 120 bulls was £22 12s. 9d. Mr. J. Searby had the best individual average, disposing of eight at £43 16s. 9d. apiece, and also the highest price, as Mr. J. Evens secured the champion, Croft Son of Violet (4781), by Croft Aubourne (4325), out of Stenigot Violet 2nd, by Stenigot Knight 2nd (527) at 150gs. Mr. H. C. Raithby, Mablethorpe, sold three at an average of £43 1s., Mablethorpe Honest Tom (4913), by Harrowby (2908) being purchased by Mr. J. F. Locking, Clee, at 52gs.; while Mr. W. Chatterton's five very young calves from Hallington averaged £33 3s. 7d., one of them by Guardian (4397) going to Mr. J. Mason, Calceby, at 61gs. Mr. S. B. Carnley deserves every credit for the success of these sales at Alford, which have developed to a wonderful degree since he took them under his management.

1908. In *1908*, at the Association's sale at Lincoln, on April 23rd, 187 bulls changed hands, the average being £28 5s. 8d., a record thus being established. The highest price this year was obtained by the Exors. of the late Mr. T. B. Freshney, for Saltfleet Marshman (4958), who was knocked down to Messrs. Thompson & Bourne, Alvingham, Louth, at 180gs. This typical Lincoln Red bull was calved in May, 1906, and so was ineligible for competition at the Show held previously. He was by Saltfleet Bonus (3582), out of that famous prize-winner Ruby 2nd, by Saltfleet Eclipse (227), who was also the dam of that beautiful cow Ruby 12th, sold for 105gs. at the dispersal of the Saltfleet herd in 1906. Messrs. T. & W. Dickinson's ten averaged £36 12s. 9d., most money being given by Messrs. S. & J. H. Briggs, Saleby, for Bonby Excursionist 17th (5837), by Scampton Excursionist (4089), who was knocked down at 52gs. Red Clay, by Anderby Alford (4214), the property of Mr. T. Wallis, St. Catherine's, Lincoln, went at 60gs. to Mr. W. R. Caudwell, Holbeach Marsh; ten very nice young bulls shown by Mr. G. Marris, Kirmington, averaged £27 1s. 9d.; and eight of Mr. Reuben

Roberts' averaged £46 1s. 4d., Thimbleby Emperor (6403), by Scampton Excelsior (4085) going at 76gs. to Messrs. T. & W. Dickinson, Worlaby. Mr. G. E. Sandars only sent four bulls this year, nearly all his 1907 calves having been heifers, and these averaged £55 13s., the first prize-winner in the older bull class Scampton Invader (5609), by Keddington Ruby (1243), being purchased by Mr. Granville Sharpe, Baumber Park, at 97gs. The pick of Mr. F. Scorer's two, which had stood next Mr. Sandars' winner in the Show ring, was Bracebridge Baron 2nd (5172), by Welbourne Red Baron (3693), and he was knocked down to Mr. Reuben Roberts, Horncastle, at 68gs.

At that year's sale at Boston, the fourth, held on May 4th, 79 bulls were sold at an average of £22 4s. od., the highest price being 70gs., which Messrs. W. Dennis & Sons, Kirton, Boston, paid for Tathwell Hero, by Calceby Hebrew (2143), the property of Mr. J. E. Davy, Tathwell. Mr. J. T. Robinson, Huttoft, sold Tothby Virtuoso 13th, by Kirkby Virtuoso (294) to Mr. W. H. Ward, Carrington, at 43gs. ; and Mr. J. Tomlinson's Birthorpe Astounding and Mr. F. S. Bett's Kirton Alfonso each made 40gs., the first going to Mr. L. W. Stephenson, South Thoresby, and the other to Mr. J. Tomlinson.

It must be very gratifying for Mr. S. B. Carnley to notice the growing success of the Alford Bull Sales, which have developed from quite a modest affair into next to the Associations' Sales, at Lincoln, in April, the most important medium for the sale and purchase of Lincolnshire Red Shorthorn bulls in the county. The catalogue numbered 144 entries, and of these nine were not brought forward, 13 failed to find purchasers, and the remaining 122 were disposed of at the satisfactory average of £23 os. 5d. This was a bit better than last year, though a trifle below that of 1906. But the general average of prices was better, for there were no big figures, as in 1906, when Mr. G. E. Sandars paid 200gs. for Mr. J. Langham's Brandon Grenadier (4274), and in 1907, when Mr. J. Searby's Croft Son of Violet (4791) was sold to go into Mr. John Evens' famous dairy herd, at 150gs. This year the top price was 75gs., which Mr. J. F. Rawnsley, Candlesby, paid for a son of Guardian (4397), sent up by Mr. William Chatterton, Hallington, for whom he had taken second honours in the class for bulls under 15 months old at the Show held previous to the sales.

By the way, the public money considerably upset the judges' verdicts during the day. His conqueror was Mablethorpe Lumberer (5503), by Hallington Bilberry (3903), who was also awarded the reserve for the Championship, being afterwards sold to Mr. Wright, Holbeach Marsh, for 39gs. He belonged to Mr. H. C. Raithby, Mablethorpe, and was also one of the four winners in the group class. The fact that his owner farms under 120 acres redounds greatly to his credit, · and shows what small farmers can do if they breed on right lines. The winner in the older bull class and of Championship honours was Benniworth Herald (5153), by Benniworth 39th, (4239), shown by Mr. C. F. Bett, Benniworth, and he was afterwards sold to Mr. Harold Dodds, Donington, Spalding, at 65gs. Mr. Betts' father won the Championship at Alford four years before. The second prize-winner in the older bull class was Mr. W. Chapman's Strubby Boy IV. (5645), by Red Chief III. (4939), for whom Mr. G. Fydell Rowley, St. Neots, paid 66gs., and Messrs. J Ranby & Son obtained 68gs. from Mr. Turner, Authorpe, Harpswell, for Bilsby Indomitable (5825), while Mr. R. Chatterton's Stenigot Ruddy, by Ashby Red II. (3728) made 70gs. to Mr. B. Rowland, Wainfleet. Mr. E. J. Turton secured a nice young bull when he paid 40gs. for Mr. G. J. Brown's Tothby Express (6433), by Calceby Express (4302); and Mr. J. Wright, New Leake; Mr. Needham, Ranby; Mr R. Clayton, Littleworth; Mr. J. Todd, Kirkby Green; Mr. C. Addison, Laceby; Mr. J. Borrill, Waudby; Mr. Spencer, Folkingham; and Mr. Starting, Withington, each gave over 30gs. for useful bulls. Mr. W. Chatterton obtained both the highest individual price and the best average, the latter being £42 10s. 6d. for three handsome young bulls.

1909. The fourteenth bull sale of the Lincolnshire Red Association was held at Lincoln, on April 22nd, *1909*, when 258 animals were brought forward, while eleven only were passed unsold. The 238 bulls which were sold averaged £28 6s. 6d., which upset the record set up last year, when 187 averaged £28 5s. 8d. At the Show held previous to the sale the four prizes in the older class went to Mr. Measures' Dunsby Red 3rd (6017), by Sausthorpe Red 12th (4552), Mr. Sandars' Scampton Jupiter (6332) and Judas (6325), and Mr. Kirke's Ravendale 3rd (6272), with Captain Grantham's Keal Skipper (6135) reserve. The winner of the younger class was Mr. Sandars' Scampton Justinian (6335), while next in order stood Messrs. Robinson's Anderby Pilot (5793), Mr. Byron's Normanby Solus (6219) and Mr. Sandars' Scampton Julia (6330), Messrs. Farrow's Strubby Hebrew 2nd (6371), being reserve. The championship went to Dunsby Red 3rd, with Scampton Jupiter reserved. The champion was of nice quality and very even in his lines, and competition was keen for him when he came into the sale ring, Mr. F. B. Wilkinson, Edwinstowe, eventually securing him for 165gs., after a tussle with Mr. John Evens and Mr. T. Wallis. Mr. Measures' two bulls averaged £101 17s. Mr. Sandars' ten bulls averaged £65 6s. 2d., most money being given for the winner in the younger class, Justinian (6335), who was sold to Mr. John Searby, Croft, at 100gs. He is by Brandon Grenadier (4274), and a most promising youngster; Jupiter (6332), a very massive, heavy-fleshed bull by Keddington Ruby (1243), went to Mr. L. Hobbs, London, at 90gs., and Judas (6325) and Julius (6330), both by Brandon Grenadier (4274), made 90gs. and 85gs. respectively, to Mr. Wallis, Lincoln, and Mr. Turner, Oadby. Messrs. Allerdice, Liverpool, paid 80gs. for Mr. Byron's Normanby Solus (6219), by Croft Marvel (3829). The following were the highest prices :—

	Gns.
Mr. J. H. Gaunt's Brandon Grenadier 3rd (5786)—H. J. Walter ..	40
Mr. F. Scorer's Bracebridge Herald 1st (5858)—W. Wright.. ..	40
Mr. W. Drake's Tealby Ranger (6394)—J. T. Tweed..	47
Colonel C. A. Swan's Sausthorpe Red 21st (6318)—R. Mowbray	40
Mr. G. E. Sandars' Scampton Jupiter (6332)—L. Hobbs	90
Mr. G. E. Sandars' Judas (6325)—T. Wallis	90
Mr. G. E. Sandars' Scampton Juvenalis (6337)—H. Jackson ..	52
Mr. G. E. Sandars' Scampton Justinian (6335)—J. Searby	100
Mr. G. E. Sandars' Scampton Julius (6330)—A. Turner ..	85
Mr. G. E. Sandars' Scampton Jude (6326)—S. E. Dean & Sons ..	44
Mr. G. E. Sandars' Scampton Juba (6324)—B. Simons	68
Messrs. T. & W. Dickinson's Bonby Excursionist 24th (5844)—	
Messrs. J. N. Robinson & Sons	40
Mr. R. Chatterton's Stenigot Comet (6360)—Mr. Willows	42
Mr. R. Chatterton's Stenigot Neptune (6364)—J. A. Milligan ..	40
Mr. J. Byron's Normanby Solus (621)—Messrs. Allerdice	80
Mr. E. Abraham's Otby Guardian (6242)—Mr. Hoyles	40
Mr. E. Abraham's Otby Grenadier (6241)—Mr. Cabourne	40
Mr. H. Abraham's Walesby King (6459) —Mr. Bingham	40
Mr. W. B. Swallow's Horkstow Meteor (6111)—J. C. Mountain ..	50
Messrs. J. N. Robinson's Anderby Pilo (5793)—Mr. Crompton ..	50
Captain E. M. Grantham's Keal Skipper (6135)—Colonel Ramsden	41
Mr. H. Kirke's Ravendale 3rd (6272)—R. J. Epton..	54
Mr. J. W. Measures' Dunsby Red 3rd (6017)—F. B. Wilkinson ..	165
Mr. J. L. Hodson's Wainfleet Baron 1st (6456)—W. B. Eve	46
Mr. A. F. Nalder's Redland Brutus (6280)—L. Hobbs	40

238 Bulls averaged £28 6s. 6d. ; total, £6,742 1s. 6d.

The ninth annual sale of Lincolnshire Red Shorthorn bulls took place at Alford on November 9th, 1909. The catalogue contained 192 entries, and 165 lots were sold at an average of £20 10s. 3d., a very satisfactory one.

Mr. W. J. Atkinson and Mr. Peter Dunn acted as judges at the show, awarding the prizes in the class for the best Bull under fifteen months to Mr. W. Chapman's Strubby Boy 5th (6368), Mr. C. F. Bett's Benniworth Herald 2nd, and Mr. G. J. Brown's Tothby Bonus (7208) ; and in the class for Bulls over 15 months to Mr. E. H. Cartwright's Nonpareil Bonus (6216), and Messrs. J. W. Farrow & Sons' Strubby Hebrew 2nd (6371) ; while in the class for Calves under nine months old the prizes went to Messrs. J. Ranby's Bilsby Dreadnought 2nd, Messrs. S. & J. H. Briggs' Saleby Excursionist 1st (7101) and Mr. S. Crawley's Showman. Mr. G. J. Brown showed the best group of four bulls, Messrs. J. W. Farrow & Sons taking the second prize ; and the championship of the show was awarded to Mr. E. H. Cartwright's Nonpareil Bonus (6211), with Messrs. Farrow's Strubby Hebrew 2nd (6371) reserve number. The champion was a heavy-fleshed and nicely-topped bull, but has not the best of hind legs ; he is by Knockabout (4899), and is a member of the famous Nonpareil family, and he was sold later on to Mr. E. Ashley at 60gs., who had stalled off Mr. B. Rowlands' attack ; the reserve for the premier honours of the show being sold to Mr. Kelk at 56gs. But most money was forthcoming for Mr. G. E. Sandars' Scampton Judge (6327), undecorated in the show ring, who was sold to Mr. W. A. Harrison at 70gs., after a spirited duel with Mr. W. Searby. He is a beautiful, dark red bull by Brandon Grenadier (7274) (champion at Alford in 1906, and sold to Mr. Sandars then for 200gs.), and is out of a Keddington Ruby (1243) cow. Mr. H. C. Raithby, Messrs. J. Ranby & Son, and Mr. G. J. Brown each obtained 40gs. for bulls.

The highest prices of the sales are as follows :— Gs.

	Gs.
J. W. Farrow & Sons' Strubby Hebrew 2nd (6371)—Mr. Kelk	56
H. C. Raithby's Mablethorpe Red Chief (6193)—B. Simons	40
G. E. Sandars' Scampton Judge (6327)—W. A. Harrison	70
E. H. Cartwright's Nonpareil Bonus (6211)—E. Ashley	60
J. Ranby & Son's Bilsby Indomitable (5825)—Lord Willoughby de Eresby	40
G. J. Brown's Tothby Dragoon 11th (7215)—J. Riggall	40

1910. The fifteenth annual sale of Lincolnshire Red Shorthorn bulls, held under the auspices of the Lincolnshire Red Shorthorn Association, took place at Lincoln, on April 28th, *1910*, when 313 animals were catalogued.

At the show previous to the sale twenty-eight bulls were entered in the older class and forty in the younger. The four prizes in the former fell to Mr. George Marris' Kirmington Forester 13th (6952), Mr. W. H. Smyth's Elkington Marshman 1st (6778), Mr. E. Abraham's Otby Hercules (7038), and Mr. G. E. Sandars' Scampton King of Trumps (7125). In the young bull class the awards went to Mr. G. E. Sandars' King of the Valley (7123), and King of Hearts (7121), Mr. Percy Hensman's Fulletby Friend (6803) and Capt. C. L. Prior's Strubby Boy 6th (7160).

- The championship of the show was awarded to Mr. Geo. Marris' Kirmington Forester 13th (6952), subsequently sold to Mr. John Evens for 82gs., the reserve being Mr. G. E. Sandars' Scampton King of the Valley (7123) knocked down to Mr. F. B. Wilkinson at 65gs.

Earl Fitzwilliam's Wentworth Earl (7248) made the highest price of the day, being sold to Capt. G. B. Portman, Healing Manor, at 112gs. Mr. G. E. Sandars obtained the best average of £44 1s. 9d. for ten very level bulls.

The supply of bulls exceeded the demand, and a heavy rain came on soon after the commencement of the sale, quickly thinning the gathering round the two rings, so that, although the quality of the animals shown was fully up to the average, prices ruled much lower than usual. The bulls were sold simultaneously in two rings, and the highest prices were as follows :—

RING 1.	Gs.
Mr. Percy Hensman's Fulletby Friend (6803)—Lord Heneage	42
Mr. G. E. Sandars' Scampton Kingdom (7118)—J. W. Belton	42
Mr. G. E. Sandars' King of Trumps (7125)—E. Dove	48
Mr. G. E. Sandars' King of the Rubies (7122)—F. Bourne	50
Mr. G. E. Sandars' King of the Valley (7123)—F. B. Wilkinson	65
Mr. G. E. Sandars' King of Hearts (7121)—L. W. Stephenson	70
Mr. G. E Sandars' King Hal (7119)—W. Bowser	40
Mr. G. E. Sandars' King Post (7126)—J. Ward	40
Messrs. T. & W. Dickinson's Emperor (6598)—E. Abraham	40
Mr. J. C. Mountain's Welbourne Artist (6466)—R. Mowbray	46
Mr. Geo. Marris' Kirmington Forester 13th (6952)—John Evens	82
Mr. Edward Abraham's Otby Hercules (7038)—John Langham	68
Messrs. J. N. Robinson & Son's Sea Dog (6545)—Mr. Wygram	40
Messrs. J. N. Robinson & Son's Coastguard (6529)—J. W. Ward	40
Earl Fitzwilliam's Wentworth Earl (7248)—Capt. G. B. Portman	112
Mr. Edwin Bourne's Ruby King 2nd (7094)—A. P. Brandt	48
Mr. A. P. Brandt's Bletchingley Apollo (5831)—A. Parkinson	45
Mr. C. W. Tindall's Auctioneer (6560)—G. Mowbray	51

Mr. G. E. Sandars has probably never put forward a more level and sorty lot than he did this year, and, as is almost invariably the case, they sold better than anything in the catalogue. Each of the four he was permitted to show received a decoration, Scampton King of Trumps (7125), afterwards sold to Mr. E. Dore, Bonby, taking the fourth prize for older Bulls, and Scampton Kingdom (7118), Mr. J. W. Belton's purchase at 42gs., a h.c. ticket, while the first two prizes for younger Bulls were awarded to Scampton King of the Valley (7123) and Scampton King of Hearts (7121), the former also being runner-up for the championship of the show. These two bulls were respectively purchased by Mr. F. B. Wilkinson at 65gs., and Mr. L. W. Stephenson at 70gs. ; Scampton King of the Rubies (7122) also going to Mr. F. Browne at 50gs., while Scampton King Hal (7119) and Scampton King Post (7126) each made 40gs. respectively to Mr. W. Bowser and Mr. J. Ward. All the ten Scampton bulls were by Brandon Grenadier (4274), and with the exception of Scampton King of the Valley were out of Keddington Ruby (1243) cows, and their average £48 1s. 9d., was the best of the day. The bulls sent up by Messrs. T. & W. Dickinson are always in demand, and this year found ready customers.

As stated last week, top price was given by Capt. the Hon. G. B. Portman, Healing Manor, who paid 112gs. for Earl Fitzwilliam's Wentworth Earl (7248), after a tussle with Mr. Robert Chatterton and Capt. C. L. Prior. This bull, which was calved in July, 1908, took a third prize at the county show last year, and is by Cropwell Dainty (4341), out of Cropwell Pride 4th, of Mr. Marriott's blood, but bred at Wentworth.

A summary of the sales held by the Lincolnshire Red Shorthorn Association at Lincoln up to date is appended :—

Year.	No. Sold.	Highest Price. gs.	Average. £	s.	d.	Total. £	s.	d.
1896	140	50	23	9	5	3,285	19	6
1897	225	60	20	10	11	4,521	16	6
1898	193	62	22	5	3	4,297	2	6
1899	147	81	27	10	11	4,049	6	6
1900	196	105	23	18	6	4,689	16	6
1901	164	85	25	10	11	4,190	0	6
1902	169	60	25	4	4	4,261	19	0
1903	189	65	26	5	3	4,963	19	0
1904	219	130	25	4	5	5,523	10	6
1905	187	100	24	11	3	4,593	5	0
1906	166	305	27	10	9	4,571	14	0
1907	210	140	26	2	11	5,490	19	6
1908	187	180	28	5	8	5,289	7	6
1909	236	165	28	10	4	6,729	19	6
1910	277	112	25	5	7	7,002	10	8
15 Sales	2905	305	£25	5	9	£73,461	6	8

G

The tenth annual sale of Lincoln Red Shorthorns held in connection with Alford Fair took place on November 20th, 1910. The weather was favourable, for though cold it was bright and clear. A catalogue of 203 animals was presented, and 190 came forward. So good was trade that only twelve were passed, 178 finding purchasers at prices ranging up to 100gs.

The show of bulls was a distinct improvement on the earlier fixtures, chiefly through greater care in bringing them out. In recent years there has been a decided improvement in the form in which Lincoln Reds are shown, and this is spreading to the larger circle interested in the bull sales. At the same time all breeders have not taken to heart the lessons to be gathered from previous sales. Many of the young bulls forward would undoubtedly have made better prices if they had been fuller of flesh.

In the class for bulls not exceeding fifteen months Mr. W. Chapman took first prize with Strubby Boy 8th (7162), a straight, thick-fleshed bull, and second and third prizes went to Messrs. J. W. Farrow & Sons for bulls by Red Chief 3rd (4939). For bulls over fifteen months Mr. E. H. Cartwright had a very nice bull in the first place. This is Keddington Grange 2nd (6928), by Grange Prince (4843). The only fault that could be found with him is that he had a little too much white on his underline, and might have been a trifle neater behind the shoulders, but he is a level, wealthy, even-fleshed bull, very full of hair and brought out in excellent form. Second honours went to Mr. C. F. Bett for Hainton Beau, a compact, dark red, but lacking the wealth of flesh of the first. Mr. W. Thorn-alley's Ashby Admiral was third. In the calf class Mr. L. Coppin won first honours with a calf by Strubby Under-Porter 2nd (6379). He is a smart, level calf; and next him came Messrs. C. & R. Brooks' calf by Bonby Excursionist 14th (5834). Mr. W. G. Smyth had third prize for the growthy Mablethorpe Ragged Boy 4th, which looks like finishing a big bull. This youngster afterwards made the second highest price of the day. Messrs. Briggs won first for groups, and the Keddington bull easily won championship honours.

The sale started with the champion bull, which was put in by Mr. J. G. Williams' agent, Mr. H. W. Bishop, at 60gs., and quickly ran up to 100gs., Mr. Bishop again being the bidder. A large number of the bulls offered were young, and in the circumstances an average of over 22gs. was satisfactory, but just towards the end there seemed to be a slackening in the trade. A hundred and seventy-eight head totalled £4,179 10s. 6d., an average of £23 9s. 7d. The following were the highest prices :—

	Gns.
Mr. J. P. L. Hodson's Wainfleet Gladiator—T. P. Paul, Wrangle Hall	46
Mrs. Massingberd-Mundy's Ormsby Heretic—R. Morley, Leadenham	40
Mr. Levi Coppin's Red—Mr. Gaunt, Wispington	41
Mr. E. H. Cartwright's Keddington Grange 2nd (6928)—J. G. Williams, Tring	100
Mr. J. Tomlinson's Birthorpe Brilliant—H. Dodds, Donington	52
Mr. T. H .B. Freshney's Saltfleet Wonder (7112)—G. F. Sleight, Weelsby	52
Mr. G. J. Brown's Tothby Dragoon 17th—C. Fieldsend, Kirmond	40
Mr. W. G. Smyth's Elkington Marshman 7th, c. 1910 (3rd prize)—J. G. Williams	70

The following is a summary of the sales held at Alford under the new organization :—

Year.	Bulls offered.	Bulls sold.	Average.
1901	76	56	15¼ Gs.
1902	84	56	20 ,,
1903	127	95	16½ ,,
1904	160	120	19 ,,
1905	115	86	21½ ,,

Year.	Bulls sold.	Highest price. Gs.	Average. £ s. d.	Total. £ s. d.
1906	104	200	25 18 7	2,696 14 0
1907	120	150	22 12 9	2,716 17 6
1908	122	150	23 0 5	2,808 15 6
1909	165	70	20 10 3	3,549 10 6
1910	178	100	23 8 4	4,168 10 0

PRIVATE AUCTION SALES.

The increasing number of private auction sales show how great is the demand for the best blood in the breed, and that the prices paid must be remunerative in the extreme. Wherever there is a draft or dispersal sale of a herd of note, there one is sure to see an average that compares most favourably with other Shorthorn sales, and that is superior to those in most breeds. And this in spite of the fact that no sensational prices are made by one or two animals, for a uniform average runs through the catalogue, and the figures paid are such as most tenant-farmers can afford.

1900. The first private sale that I shall deal with is that held by Messrs. S. & J. Crawley, at Hemington, Oundle, in Northamptonshire, on October 4th, *1900*, when 61 head were sold at an average of £25 19s. 7d. The thirty-one cows and heifers, with calf or in-calf averaged £31, and seventeen maiden heifers £16 3s., while seven yearling bulls averaged £30 12s., four bull calves £17 3s. 10d., and two stud bulls £32 0s. 9d. Of the females Thurlby Duchess, of Mr. Wallis Wells' breed, went to Mr. J. G. Williams, Pendley Manor, Tring, at 41gs. and her heifer calf by Baron Ormsby 3rd (26) was knocked down to Mr. John Gray, East Norton, at 31gs. Thurlby Duchess was the dam of Bumper 2nd (1793) for whom Mr. W. B. Swallow, Wootton Lawn, Ulceby, paid Messrs. Crawley 105gs. at Lincoln Fair in the spring of that year. The pick of the two-year-old heifers, Edna May, by Baron Ormsby 3rd (26), from Thurlby Princess, also went to Mr. Williams at 25gs. The yearling bull Irish Lord (2199), by Lord Chancellor (1606) out of Well Beauty 3rd was bought by Mr. Langton, Yarborough, at 38gs. Marmion, another yearling by Baron Ormsby 3rd (26) out of Thurlby Princess, fell to Messrs. R. & R. Chatterton, Stenigot, at 42gs. Mr. C. W. Tindall, Wainfleet Hall, gave 32gs. for the bull calf by Lord Chancellor (1606) from Well Rose 2nd ; and Mr. L. W. Stephenson, South Thoresby, paid 36gs. for the stud bull Baron Ormsby 3rd (26), by Stamp End (247), from a cow combining the blood of Mr. Needham's Legbourne herd with that of Mr. Oliver, of Eresby.

Another sale held this year was at Fledborough House, near Newark, when 41 head, the property of Mr. C. H. Stafford, were disposed of at an average of £35 5s. 10d. The pick of the cows and heifers, with calves or in-calf, was Withern Violet, by Marshman (180) from Withern Primrose, of Mr. Wallis Wells' strain, who went into the famous milking herd of Mr. John Evens, Burton, at 41gs. ; the 28 lots averaging £37 18s. 7d., while 10 maiden heifers averaged £21 7s. 2d. One of these, a sweet yearling heifer, Lady Sarah by name, and by Calceby Striver (663), out of Croft Ruby, was bought by Colonel Thorpe, Coddington Hall, Newark, at 57gs. The three bulls averaged £29 15s., that wonderful sire Benniworth 4th (629), by Keddington (151), out of a Benniworth Kelstern (28) dam, who was bred by Mr. T. Bett, Benniworth, and was then four years of age, going to Mr. E. H. Cartwright, Keddington, at 50gs.

1901. Mr. John Searby, Croft, Wainfleet, had a draft sale, on September 18th, *1901*, when the highest prices obtained were 30gs. which Mr. A. Marshall, Humberstone, paid for Octavius (2280), by Keddington Ruby (1243), and 37gs. which Mr. R. S. Hand gave for Keal Empress, by Lord Rosebery (839) and her bull calf by Lord Kitchener (1929). At the same sale Mr. Searby sold Keddington Ruby (1243) to Mr. G. E. Sandars, Scampton, privately, in whose herd he proved a most wonderful sire, 40 of his sons averaged over 60gs. at the Lincolnshire Red Shorthorn Association's sales held at Lincoln each April.

The same year, on September 26th, the herd of the late Mr. W. Scorer, Sudbrook, was disposed of, cows making up to 27gs., heifers up to 24½gs., and bulls up to 25½gs. ; and on October 10th, Mr. W. Hornsby's herd at Burwell Park, Louth, was also dispersed, the 55 lots averaging £18 0s. 3d. with 28½gs. as the highest price for a cow and bull calf.

A draft from the herd of Mr. J. Langham sold at Hough Grange and numbering 65 head averaged £17 7s. 6d. a bull making 36gs. and a cow 28gs. ; and on September 26th, 77 head selected from the herd belonging to Mr. W. J. Atkinson, Weston St. Mary, averaged £21. The cow Weston Red Plush, by Lord Knightley (C.H.B. 67,364) made 40gs. to Mr. R. Clayton, Littleworth, and Mr. John Evens, Burton, paid 35gs. for her bull calf Weston Hero 2nd (2703), by Red Monarch (C.H.B. 77,605). The 12 yearling heifers had an average of £17 18s. 9d. and the 9 bulls £25 11s., Mr. W. Ward giving 40gs. for one.

But the best private sale that year was at Stenigot on October 17th, when a choice selection of 89 animals from Messrs. R. & R. Chatterton's well-known herd were sold at an average of £35 10s. 10d. The 42 cows and calves averaged £41 6s. 6d., Stenigot Violet 2nd, by Stenigot Knight 2nd (527) out of Stenigot Violet, by Comet (79) going to Mr. J. Searby, Croft, at 45gs. She was a wonderful prize-winner for Messrs. Chatterton, and she continued her show-yard successes after she changed hands. A bull calf by Sirdar (1676) from Stenigot Bloom, by Candidate (56) made 40gs. to Mr. R. M. Knowles, Colston Bassett Hall, Notts. ; and the two-year-old heifer Stenigot Daisy VI., by Wrangler (C.H.B. 71,901) from Stenigot Daisy II., by Comet (79) was sold to Mr. W. J. Atkinson, for 52gs. ; while Mr. G. J. Brown, Tothby and Lord Heneage, Hainton Hall, gave 47gs. and, 45gs. respectively for Stenigot Daisy VII., by Wolseley (1436) and Stenigot

Pansy VII. by the same sire. The 32 heifers averaged £27 14s. 6d. and the 15 bulls £36 0s. 3d. Foremost among these latter was the prize-winner and successful sire Red Chief (2611), by Sirdar (1676) out of Stenigot Red Rose III., by Crœsus (85), her dam Red Rose, by Prince Gwynne III. (C.H.B. 43,799), and her grand-dam by Hyllus (144), who was knocked down to Mr. T. Bett, Benniworth, after a spirited duel with Mr. W. Wright, Nocton. The young bull Royal Herald, also by Sirdar, was bought by Mr. W. Nainby, Thorganby Hall, at 55gs.

1902. In *1902* the herd belonging to Mr. Wallis Wells, Withern, was dispersed on October 2nd, at a total sum of £2,177. The fifty-five cows and heifers averaged £18 2s., and twenty-five calves £9 17s. 2d., the top price for cows being 27½gs., while a pair of maiden heifers made 55gs. Mr. Walter Wingfield's herd at Driby Manor, was dispersed on October 7th and realized the sum of £1,013 10s. 3d. for 71 animals. The cows averaged £19 15s. 9d., calves £9 4s. 1d., two-year-old heifers £20 7s. 3d., yearling heifers £12 3s. 7d., and the stud bull, General Buller (2206) made 26½gs. The herd belonging to Mr. G. A. Brown, Highfield, Cadwell, was sold two days later a beautiful lot of cows deserving a better average than £23 6s. Mr. T. Diggle's herd at Ewerby was also sold on October 16th, when reasonable prices were obtained.

1903. Owing to the land being required by his hunters, Sir Robert Wilmot, Bart., the Master of the Berks and Bucks Staghounds, disposed of his herd at Binfield Grove, Bracknel, in *1903*, Messrs. R. & R. Chatterton paying 42gs. for the two-year-old Royal winner, Binfield Brilliant, a handsome young bull, while Mr. J. Marriott gave 31½gs. for one of the cows, and Messrs. Purser bought a number at prices ranging up to 28½gs., and Lady Errington, Lord George Pratt, Sir Charles Bull, and Colonel Van-de-Weyer being also purchasers.

There were no other private auction sales of any importance that year, but Mr. E. H. Cartwright, Keddington, Mr. J. Evens, Burton, Messrs. R. & R. Chatterton, Stenigot, Messrs. S. E. Dean & Sons, Dowsby Hall, and Mr. C. W. Tindall, Wainfleet Hall, sold a great number of animals privately to go abroad, as well as to home breeders. Mr. J. K. Hill, of Bloemfontein, and Mr. Burbidge, of Graff Reinet, Cape Colony, made several purchases from the Stenigot herd to go to South

Africa ; the latter gentleman, a born colonial of 55 years' experience, who had been touring through England and Scotland with a view to buying cattle for the country, stating that he found the Lincoln Reds more suited to its requirements than any other breed. Both Messrs. Dean and Mr. J. Marriott, Cropwell Butler, sent animals to South Africa, and the former gentleman also made shipments to South America ; while Mr. C. W. Tindall sent a number of his purchases all over England and to the Continent.

1904. In *1904* there were several notable private sales, the first being on September 21st, when a draft numbering 104 head from the Northborough Castle herd belonging to Mr. Everett King, were disposed of at an average of £18 14s. 6d. Here the 17 cows with their calves averaged £26 12s. 8d., Mr. Sheldon, Leicester, giving 32½gs. for Wainfleet Maud, by King's Bendigo (2245) and 32gs. for Cadwell Bess, by Sunbeam (1370). The cows in calf, numbering 26, averaged £22 13s. 10d., most money being 30gs., which Mr. T. Berry, Thorney, paid for Northborough Bot, by Benniworth VII. (1458) ; the 19 heifers in calf averaged £18 12s. 9d. ; the 21 yearling heifers, £13 5s. ; seven calves, £5 13s. 3d. ; and 13 bulls and bull calves, £17 15s. ; the top price being paid for the Royal and Lincolnshire show winner, Northborough Cromwell IV. (2587), then four years of age, who was sold to Mr. S. Crawley, Hemington, Oundle, at 45gs. He was by Baron Ormsby III. (26) and came of Mr. Wallis Wells' strains on his dam's side.

But the most successful home sale was that of a selection of 65 animals from the herd of Mr. W. J. Atkinson, Weston St. Mary, the average for which was £27 4s. The 14 in-calvers averaged £35 14s., most money being forthcoming for Choice Juliet 2nd who was knocked down to Messrs. S. E. Dean & Sons, Dowsby Hall, Bourne, after a duel with Mr. J. W. Measures, Dunsby, at 80gs. She was by Red Monarch (C.H.B. 77,605) from Princess of Fashion, by Hindlip Boy (C.H.B. 68,775), and was in calf to Imperial Favourite (C.H.B. 86,233), for whom Messrs. Dean paid 700gs. as a calf at the late Mr. W. S. Marr's sale in *1903*. Mr. T. Stokes, Warmington, also gave 50gs. for Royal Lady 2nd, by Red Monarch out of Christmas Joy, by Hindlip's Boy, also in calf to Imperial Favourite (C.H.B. 86,233). Foremost among the 32 cows and calves, whose average was £21 14s. 5d., stood Saltfleet Favourite and her handsome bull calf Western Somercotes (3703), the former going into Mr. John Evens' famous milking

herd at Burton, near Lincoln, at 60gs., and the latter to Mr.
R. J. Epton, Wainfleet, at 46gs.   This grand cow, who won
for Mr. Evens both at the Royal and the Lincolnshire Shows,
was by Grandad (1561) from a cow by Nonsuch (1292), who
was out of Nonpareil, by Hyllus (149), and her calf, a winner
at the Lincolnshire Show that year at Grimsby, was by Saltfleet Combatant (2319).   Nine yearling heifers were sold at
an average of £20 8s. 4d., and nine bull calves at an average
of £29 5s. 8d., one of them being Weston Bob 2nd, by Royal
Crest (C.H.B. 82,149), who was knocked down to Mr. T. W.
Atkinson, Greatford, at 52gs.   But the top price of the day
was obtained for Weston Monarch 2nd (3144), a noted prize-
winner, by Red Monarch (C.H.B. 77,605), out of Western
Charm, by Pippin's Pride (C.H.B. 71,157), who was sold to
Mr. S. Crawley, Hemington, at 120gs.

The next day a draft from the herd belonging to Mr. G.
Freir, Deeping St. Nicholas, was offered, 70 lots obtained an
average of £23 9s. 11d.   The 30 cows and calves averaged
£30 18s. 8d., one of them Choice Juliet, own sister to Mr.
Atkinson's high-priced heifer of the previous day, being
knocked down to the latter gentleman at 50gs., while her cow
calf by Buscot Loyalty (C.H.B. 80,591) went at 22gs. to Mr.
Clayton, Deeping.   Five in-calf heifers averaged £25 6s. 1d. ;
nine two-year-old heifers £18 8s. 8d. ;   14 yearling heifers
£15 11s. 2d. ;   and 12 bulls £17 3s. 10d.

Mr. C. O. Parr's herd was dispersed at the Grange Farm,
Well, on October 4th at rather low figures ;   and the late Mr.
T. Stafford's at Marnham Grange, Newark, on the 11th of the
same month, when nine cows and calves averaged £29 1s.,
and 25 in-calvers £17 9s., maiden heifers making up to 20gs. ;
while two days later the herd belonging to the late Mr. T.
Frearson, Stixwould Abbey, was sold, six cows and calves
bringing an average of £25 2s. 3d., and 20 in-calvers £16 3s.,
while first-calf heifers made up to 24gs.   The bull Stenigot
Ruby Chief (3621), by Red Chief (9611) was sold to Mr. Frearson, Barkwith, at 45gs.

Another dispersal sale took place at Brough-on-Bain, on
October 27th, this being the late Mr. J. Walker's herd.   Mr.
G. E. Sandars, Scampton, bought young heifers at prices
running up to 26½gs., and both Lord Heneage and Mr. J. Evens
paid a little less for some ;   while bull calves made up to 20gs.
and heifer calves up to 18gs., the pick of the former going to
Mr. J. Phillipson, Hainton.

1905. The three most important private auction sales in *1905* were those held at Mere Hall, Lincoln, on April 26th, when Messrs. S. E. Dean & Sons, Dowsby Hall, disposed of their herd of Lincoln Reds; on September 21st, when a selection from the herd of Mr. J. Marriott, Cropwell Butler, Notts., was offered; and September 27th, when the late Mr. A. Smith's herd at Surfleet was dispersed. At the first of these 81 lots were sold at an average of £27 11s. 5d., the 26 cows and calves averaging £35 8s. 1d.; 10 two-year-old heifers, £30 11s. 1d.; 11 eighteen-months-old heifers, £30 14s. 8d.; 18 yearling heifers, £20 18s. 10d.; and 16 yearling bulls, £17 5s. 2d. Most of the cows were in-calf to the 700-guinea bull Imperial Favourite (C.H.B. 86,233) (afterwards sold in Buenos Ayres for 1,000gs.), Lord Wild Eyes 18th (C.H.B.) and Jock Scot (C.H.B.), and most money for cows was given by Mr. J. Searby, Croft, who paid 38gs. for Dowsby Sharman Oliver 3rd, by Dowsby Waterloo Oliver 2nd (C.H.B. 72,372), while Mr. J. Langham paid 40gs. for the bull calf by Imperial Favourite (C.H.B. 86,233) out of Dowsby Belle, by Virtuoso (C.H.B. 69,763). Two pretty yearling heifers made 30gs. each, and the yearling bull Dowsby Virtuoso 28th (4363), by Virtuoso (C.H.B. 69,763) made 33gs. to Mr. G. Freir.

Only moderate prices were obtained at the Cropwell Butler sale, where 61 head averaged £15 12s. 7d., and Messrs. R. & R. Chatterton secured Cropwell Miss Wainfleet, by Bigby Roughcoat (1461), at 35gs., and the yearling bull Cropwell Lawyer (3841) made 36gs. to Mr. Skipton, Nottingham.

At the Surfleet dispersal sale more money was obtainable, and the average of £20 8s. 8d. for 83 lots was quite a good one. The 64 cows and heifers and calves averaged £20 8s. 6d., Mr. C. T. Garland, Moreton Morrell, Warwick, who made several purchases, paying 42gs. for Surfleet Daisy, by Lord Fairfax XIII. (1925), and 46gs. for the two-year-old heifer, Surfleet Dorothy, by Scampton Bloodsucker (3043), a prize-winner. Thirteen yearling heifers averaged £20 0s. 7d., Surfleet Ivy, by Surfleet Baron (2666) making 37gs. to Mr. Topham, Thorney; while the six bulls averaged £21 8s. 8d., Mr. Garland giving 39gs. for Surfleet Baronet (4127), own brother to Mr. Topham's purchase.

1906. The auctioneer's hammer was busy in *1906*, and there were several interesting home sales, the first being the dispersal of the late Mr. T. B. Freshney's famous herd at Salt-fleet, on March 20th. Here 56 lots were sold at an average of

£37 8s., twelve of the noted Ruby family averaging £37 8s.
Ruby 6th, Bigby (319) out of Ruby 1st, whose dam, Red Ruby,
made the highest price at the dispersal sale of Mr. Richard
Mason's herd at Keddington, in 1872, where seven Ruby
females averaged 63gs., was bought by Mr. Richard Chatterton,
Stenigot, at 40gs., Mr. J. Crawley, Church Lawford, Rugby,
giving 29gs. for her baby bull calf by Saltfleet Bonus (3582).
Ruby 12th, by that fine bull Benniworth 4th (629) from Ruby
2nd, by Saltfleet Eclipse (227) out of Ruby 1st, made the top
price of the day, being knocked down to Mr. E. Bourne, Louth,
at 106gs.   Mr. Robert Chatterton was runner-up ; and Ruby
12th greatly distinguished herself at the Royal Show at
Lincoln the following year.   Her dam. Ruby 2nd, was bought
by Mr. E. H. Cartwright, at 50gs., after a contest with Mr.
J. W. Measures, Dunsby Manor, Bourne ; and her calf Saltfleet Imperialist (4549), by Imperial Favourite (C.H.B. 86,233)
went at 35gs. to Mr. R. Holden.   Mr. Cartwright bid well for
Saltfleet Bonus (3582), champion shorthorn at the Lincolnshire Show the year before, but Mr. J. Ward, Withcall, secured
him at 50gs.   He was by Red Monarch (C.H.B. 77,605) from
Western Nonpareil, by Lord Knightley (170).

On April 4th, there was a sale of surplus cattle from Mr.
Reuben Roberts' farm, at Digby, in consequence of his having
let the farm.   Here cows made up to 20gs., three-year-old
heifers up to 21gs., two-year-olds up to 14¾gs., and yearlings
10½gs., while the pick of the yearling bulls realized 34gs.   The
cows made up to £23 15s. at the dispersal sale of the late Mr.
H. E. Davy's herd at Croxby Hall, on April 20th ; and at a
sale of surplus stock from Mr. G. Marris' herd at Kirmington,
on April 21st cows made up to £21 10s., two-year-old heifers
£14 7s. 6d., yearlings £11 5s., and bulls £24 5s.   The dispersal of Mr. H. T. Hincks' herd of dairy cattle at Wigston
Hall, near Leicester, was on April 30th, Messrs. Fletcher &
Andrews, of Horninghold, Uppingham, bought the two best
cows, Clara, by The Bard (986), a prize-winner afterwards,
going very cheaply at 26gs., and Lassie, by Lord Gwynne
(1927), at 20gs., another bargain.   One of two bulls, P. P., by
President (C.H.B. 81,881) was knocked down to Earl Manvers
at 40gs.

Mr. Everett King's surplus sale at Church Farm, Cotterstock, Northants, took place on September 19th, moderate
prices being obtained.   But Mr. Moore took Wainfleet Maud
2nd into Nottinghamshire at 37gs., and he also bought the
yearling heifers, Northboro' Lady and Northboro' Ruby, at

30gs. and 23gs. A sale of Lincoln Reds belonging to Mr. Robert Mowbray was held at Gosberton, on September 24th, when 83 head averaged £13 13s. 1d., the pick of the two-year-old heifers being knocked down to Messrs. Lumby & Son, Donington, at 27gs., while a yearling heifer by Clan Macdonald (C.H.B. 78,597) fell to Mr. W. J. Atkinson at 30gs.

There were some big, roomy cows at the sale of Mr. H. C. Tinsley's cattle at Hurn Hall, Holbeach, on September 26th, and the average for 86 head was £16 13s. 4d. The calves were a good lot and a nice colour. Mr. W. J. Atkinson paid 36gs. for Hurn Milkmaid 2nd, and he secured Hurn Ladybird 2nd at 36gs. The bull Bingham Lodge Benniworth (2752) went to Mr. E. Ashley, Godmanchester, at 31gs.

The late Mr. T. Betts' famous herd was dispersed very cheaply on October 4th, the 57 lots averaging £25 9s. 2d. ; the cows and heifers averaged £30 12s. 7d., and the bulls £26 15s. 6d. Mr. J. F. Locking, Clee, bought the cow Benniworth Cherry, by Duke (1870) at 40gs., and a heifer calf by Saltfleet Echo (3038) from Benniworth Pearl, was knocked down to Mr. J. G. Williams, Pendley Manor, Tring, at 30gs. Mr. W. J. Atkinson took the yearling bull Benniworth 39th (4239), by Saltfleet Echo (3038) to Weston St. Mary, at 50gs. The cattle were shown in rather backward condition owing to the poor pastures. But the great private sale of the year was when a choice selection from the noted Keddington herd was offered by Mr. E. H. Cartwright, on October 17th, 65 head obtaining an average of £40 9s. The best Lincoln Red breeders, both far and near, came to get some of this famous blood, and they generally had to pay what it was worth. The 34 cows and heifers with calves averaged £43 6s. 1d., the first to make big money being that famous prize-winner, Star of the Night, by Shooting Star (1674), who was bought by Mr. J. W. Ward, Withcall, while her heifer calf went to the same buyer at 63gs. This handsome little lady was by Stenigot Bloom Boy (3611), who was sold to Senor José Abasoto, to go to Argentina, at 200gs. Next, Starlight, a beautiful cow by The Count (1398) fell to Mr. J. G. Williams, Pendley Manor, Tring, at 66gs., and her bull calf by Stenigot Bloom Boy (3611) became the purchase of Mr. T. Bell, Poolham, at 41gs. Mr. J. W. Ward secured Miss Calceby 3rd, by Bigby (319) at 50gs., and Mr. Williams paid 87gs. for Nonpareil 9th, by Conisholme Boy (347), while Keddington Skipworth 5th, by Benniworth 4th (629) was another of Mr. Ward's purchases at 91gs. The two-year-old heifers were a beautiful lot, and averaged £37

17s. 9d. for 18, most money being paid for Keddington Favourite 4th, a truly beautiful heifer by the weighty Vanguard (2691), and Keddington Favourite 3rd by Conisholme Boy (347). This typical Lincoln Red was knocked down to Mr. George Marris, Kirmington House, Brocklesby, at 104gs., both Mr. Williams and Mr. Ward having bid well for her. For her new owner she won premier honours at the Royal Show at Lincoln, and again the following year at the Royal at Newcastle. Another nice two-year-old heifer, Twilight, by Vanguard (2691) from Starlight, fell to Colonel C. A. Swan, Sausthorpe Hall, at 62gs., and Lord Heneage secured two sweet heifers at 47gs. and 46gs. respectively.

Yearling heifers to the number of 11 averaged £33 13s. 11d., Mr. Williams paying 61gs. for Keddington Skipworth 7th, by Saltfleet Bonus (3582), while Keddington Vain Girl, a handsome daughter of Stenigot Bloom Boy (3611), and Keddington Lassie, fell to Mr. Robert Chatterton's bid of 76gs. The bull Keddington Baron (4881), by Saltfleet Bonus (3582) out of Keddington Baroness, by Benniworth 4th (629) was knocked down to Mr. S. B. Carnley, Alford, at 82gs.

1907.   There were a great number of private auction sales again in *1907*, the first of any interest being the dispersal of the late Mr. J. Hodgson's herd at Nocton, on March 13th, when cows made up to 25gs., heifers 19½gs., and bulls 23½gs., Mr. Pate, of Coleby, Mr. Oglesby, Hemswell, and Mr. J. Cartwright, Dunston Pillar, being the highest bidders in each case.

Mr. Joseph Bowser, of Thornton House, Frithville, sold his herd of 178 animals on March 21st, when Mr. J. Searby, Croft, gave 30gs. for a cow, Mr. G. Freir, Deeping St. Nicholas, 26gs., and Mr. C. W. Tindall, Wainfleet, 25gs., and two-year-old heifers and yearlings made up to 19gs. and 20gs. respectively.

A highly successful sale was held at Deeping St. Nicholas, on September 20th, when a very choice selection from Mr. George Freir's herd was submitted to the public, the 48 lots obtaining an average of £26 19s. The 41 females averaged £28 4s. 8d. Mr. Trueman, Whittlesea, gave 30gs. for the cow Surfleet Minnie, of Scampton Admiral (2326), and Deeping Imperial, by Imperial Favourite (C.H.B. 86,233) out of Gold Bracelet 8th, was sold to Mr. G. Marris, Kirmington, at the same figure. This cow's sire it will be remembered was bought at Mr. W. S. Marr's sale for 700gs. as a calf by Messrs. S. E. Dean & Sons, and afterwards sold by them in Buenos Ayres for the equivalent of 1,000gs. Mr. F. Money, Sleaford, and Mr. W. Ward, Quarrington, each paid 40gs. for good cows, the first getting Deeping Gold Bracelet and the other Gold Bracelet 8th, by Buscot Loyalty (C.H.B. 80,591) and Prince Beauty (C.H.B. 77,489) respectively. Mr. E. Turnor, Oadby, Leicester, gave 50gs. for the three-year-old heifer Deeping Sutton, by Stenigot Cherry King (3613) from a Bonheur (1786) dam, and Mr. T. H. Vergette, Peterborough, secured Deeping Flora 1st by Buscot Loyalty (C.H.B. 80,591) at 40gs. Mr. Turnor also secured the two beautiful yearling heifers Deeping Princess and Deeping Queen, both by Scampton Exclusion (4088) at 50gs. and 37gs. respectively.

The dispersal of Mr. W. A. Ewbank's herd at Covenham Manor, took place on September 26th, when the cows and calves averaged £31 16s. 10d., Captain C. L. Prior, Grimblethorpe Hall, taking the best at 44½gs. Mr. J. W. Ward, Withcall, bought two-year-old heifers at 30½gs. each, and yearlings at 17gs. each. The bull Hallington Neptune (3904), was sold to Mr. R. Chatterton, Stenigot, for whom he won many prizes, at 36gs. The average for 72 lots was £23 14s. 8d.

A selection from the herd belonging to Mr. S. Crawley, Church Lawford Grange, Rugby, was sold on October 4th, 42 head averaging £23 11s. 9d. The 34 females averaged £23 14s. 4d., and 8 bulls £23 0s. 8d. The cow Sea Gem, by Sirdar (1676), from Stenigot Pansy 4th, went to Mr. H. Stokes, Nassington, at 51gs., and Ladylike, by Blood Royal (2417) out of Thurlby Princess, was knocked down to Rev. W. Hopkinson, Wansford, at 40gs., Mr. W. J. Atkinson, Weston, Mr. E. Ashley, Godmanchester, and Messrs. Fletcher & Andrews, Horninghold, Uppingham, also making judicious purchases. The bull Birk's Jewel (C.H.B. 94,335), by Holker Jewel (C.H.B. 81,280), was sold to Mr. W. J. Atkinson, at 91gs., the top price of the day.

The small herd belonging to the late Mr. G. Laughton, Belchford, was dispersed on October 7th, the 38 head averaging £23 4s. 2d. The cow Ruby 4th, by Saltfleet Sappy (502) the winner of many prizes herself and the dam of Ruby 14th, another great prize-winner, was sold to Mr. J. Marriott, Cropwell Butler, Notts., at 38gs.; and Messrs. C. Hensman & Sons, Fulletby, gave 46gs. for a good two-year-old Norbury Cato (2993) heifer. A strong draft from the herd belonging to Messrs. J. W. Farrow & Sons, Strubby Manor, Alford, was disposed of on October 8th, the average for 71 head being £30 13s. 7d. The cows and calves were the features of the sale, and of these Strubby Red Lady 2nd, by Red Curly Coat

(923) made 40gs. to Mr. B. Rowland, Wainfleet, while Ivory Horn, by the same sire, went at a like figure to Mr. J. Wells, Withern. But Ivory Horn 2nd, by Ormsby Baronet 2nd (2283) out of Ivory Horn, made most money in the sale, going to Mr. J. Drewery, Donington, at 42gs., while her bull calf by Under Porter (3126) fell to the bid of Mr. W. M. Casswell, North Ormsby Hall, at 30gs. The two-years-old heifer, own sister to the above cow went at 35gs. to Mr. A. Turner, Oadby, who also paid 40gs. for the yearling Lady Cardiff 4th, by Under Porter (3126). The yearling bull Strubby Under Porter 2nd, by Under Porter was sold to Mr. J. R. Copping, at 30gs.

Good prices marked the sale of Mr. W. J. Atkinson's selection at Weston St. Mary, on October 9th, when an average of £34 2s. 8d. was obtained for 63 lots. The cows and calves sold well, the heifer calves with their dams averaging £46 5s. 7d., while the bull calves and their mothers had an average of £45 9s. 8d. Of the former, Keddington Enderby 3rd, by Scampton Artist (2328) was sold to Mr. R. Clayton, Little-worth, at 45gs. and her heifer calf, Weston Keddington, went into the Stenigot herd at 34gs. Hurn Milkmaid 3rd, by Hurn 1st (2336) went to Mr. W. S. Smith, Thorney Fen, at 41gs., and her bull calf Weston Milkman, by Bingham Lodge Benni-worth (2572) fell to Mr. F. Lumby, Donington, at 25½gs. The heifer Surfleet Royal Lady, by Buscot Loyalty (C.H.B. 80,591), was knocked down to Mr. T. H. Vergette, Boro' Fen, at 45gs., but the top price of the day was obtained for Nonpareil 16th, a handsome two-year-old heifer, by Scampton Expansion (4093) out of Nonpareil 6th, who became the property of Mr. T. W. R. Atkinson, Stamford, at 52gs. Messrs. C. Hensman & Sons, Fulletby, paid 41gs. for the Royal winner, Western Pink, by Clan Macdonald (C.H.B. 78,597) ; and the yearling bull Weston Victor, by Scampton Excelsior (4085) made 38gs. to Mr. James Rawlinson, Pinchbeck.

The cattle offered by Mr. J. Langham, Brandon Grange, on October 10th, did not make the prices they deserved, and the 65 head were worth more than £23 3s. 7d. apiece. Mr. F. Taylor, Nether Broughton, gave 50gs. for the two-year-old heifer Brandon Beauty, by Brandon Christmas Gift (3256), while the sweet yearling heifer Brandon Ruby Nonpareil by the same sire out of Nonpareil 9th, by Admiral (3) of Kedding-ton blood, was knocked down to Mr. Robert Chatterton, Stenigot, at 52gs., he also paying 48gs. for Brandon Nonpareil Ruby, by Brandon Lord Chief Justice (2425) from Nonpareil 5th, another cow of Mr. E. H. Cartwright's breeding.

A sale of the late Mr. J. H. Tyson's herd was held at Cadeby Hall, on Otcober 17th, when cows, heifers and calves averaged £24 17s. 10d., in-calvers £17 6s. 6d., two-year-old heifers 17gs., and yearlings £12 18s. 2d.

But the best sale of the year was at Stenigot on October 24th, when Mr. Robert Chatterton placed an exceedingly choice selection before the public, the merits of which the latter were not slow to appreciate. And so the average for the 66 lots was £43 10s. 4d., and the highest price one guinea short of 100gs. The cows and heifers with calves numbered 36, and their purchasers valued them at no less than £51 6s. 1d. apiece. Stenigot Gwynne 16th, by Royal Sovereign (3034) and her heifer calf by Ashby Red 2nd (3728) went respectively to Mr. G. Marris, Kirmington, at 48gs. and Mr. J. Langham, Brandon Grange, at 36gs., and Stenigot Violet 8th and her bull calf by Cornish Knight (C.H.B. 78,641) went to Mr. W. M. Casswell, North Ormsby Hall, at 30gs. and Mr. J. Berry, Thorney, at 34gs.; while Stenigot Duchess 10th and her son by Ashby Red 2nd (3728) made 31gs. to Messrs. Fletcher & Andrews, Uppingham, and 31gs. to Mr. E. H. Cartwright. Mr. A. Turner, Oadby, bought Stenigot Flower 23rd, at 40gs. Most money for two-year-old heifers was obtained for Stenigot Daisy 21st, by Red Chief (2611) from Western Daisy, who fell to Mr. George Marris' bid at 44gs., and the highest-priced yearling proved to be Stenigot Buttercup 17th, who at 31gs. became the property of Mr. J. Thompson, East Keal. The eight bulls averaged £55 15s. 7d. and the pick of the bunch was sold to go to South Africa, at 99gs. This was the Royal winner, Stenigot Duchess Beau (5632), who was calved in February, 1906, being by Stenigot Beau 2nd (4107) out of Stenigot Duchess 10th, by Sirdar (2676), and her purchaser was Mr. D. Frame, who had Mr. J. F. Locking, Clee, and Mr. E. Ashley, Godmanchester, as serious rivals. Flower Chief (4994), by Red Chief (2611), went into Derbyshire as Mr. Swanston's purchase, at 52gs., and King Cowslip (5455), by Ashby Red 2nd (3728), out of Stenigot Cowslip 2nd, by Bally Walter (23) fell to Mr. E. Ashley, Godmanchester, at 72gs., while Welcome Chief (5738), by Red Chief (2671) from Welcome Lady, by Conisholme (52), was purchased by Mr. Eustace Barlow to go to South Africa, at 62gs.

1908. Reference must be made to the sale of the late Mr Coates Sharpley's enormous herd of nearly 600 head at Tows Grange and Brackenborough Hall, in April, *1908*, for although unregistered there was no purer bred Lincoln Red herd in the county. People came from all parts of the country, and considering that the animals were ineligible for the Herd Book, the prices ruled high on both days of the sale, the cows making up to 37gs., two-year-old heifers to 31gs., yearling heifers to 18½gs., while the home-bred (non-pedigree) bulls made up to 30gs., the general average being a very good one.

At Mr. George Marris' draft sale, at Kirmington, near Brocklesby, on September 16th, the 15 cows and calves obtained an average of £32 19s. 9d., Mr. J. G. Williams, Pendley Manor, Tring, paying 50gs. for the well-known prize-winner, Keddington Favourite 4th, by Vanguard (2691), from Keddington Favourite 3rd, by Conisholme Boy (347), her handsome bull calf by Scampton Excursionist (4089) falling to Mr. R. Chatterton, Stenigot, at 34gs. Mr. W. Wood, buying for South Africa, paid 30gs. for Lady Bess 2nd, by Hallington Impressive (1442), and 43gs. for Stenigot Gwynne 16th, by Royal Sovereign (3034). The two-year-old and yearling heifers ought to have made more money, but the general average was £23 15s. 3d.

At a sale of the herd of the late Major G. Allenby Browne, Maidenwell, cows made up to 30gs., and two-year-old heifers to 26gs., while a beautiful yearling, by Bonny Boy (3758), was sold to Mr. F. B. Wilkinson, Edwinstowe, Newark, at 32gs. The sale of 60 head, the property of Mr. John Searby, Croft, Wainfleet, on October 15th, resulted in an average of £27 7s., the 52 cows and heifers selling at £27 7s. 4d. apiece, and the 8 bulls at £27 4s. 8d. The cow, Croft Rachel III., by Keddington Ruby (1243) went at 32gs. to Mr. J. G. Williams, Pendley Manor, Tring, while Mr. Watson, Burwell, paid 30gs. for her bull calf by Croft Aubourne (4325). Mr. Williams also paid 40gs. for Croft Bilsby Ruby, by Keddington Ruby (1243), and Messrs. C. Hensman & Son, Fulletby, took Croft Marvel II. a pretty yearling heifer by Croft Aubourne (4325), at 27gs. For the yearling bull Croft Violet King (5275), by Croft Aubourne (4325) out of Croft Violet, by Croft Kindly (2837), Mr. R. J. Epton, Wainfleet, paid 40gs., but the top price of the day was obtained for Croft Sittyton, by the same sire, but out of Croft Golden Ruby, by Keddington Ruby (1243), who fell to the bid of Mr. G. J. Brown, Tothby, at 56gs. Mr. T. Gaunt, Wispington, disposed of 55 head at an average of £24 4s., on October 22nd; while on October 29th, the 46 lots belonging to Mr. J. T. Cox, Foston Lodge, Leicester, obtained an average of £21 4s., the younger stock not making much money. Mr. C. H. Fleetwood-Hesketh, Stretton, Oakham, gave 38gs. for Foston Regina IV., by Foston Duke (2204), and Foston Red Rust IV. and Foston Lucy, by the same sire, went respectively to Mr. Mark Frith, Wiston Hall, at 34gs. and Capt. the Hon. G. B. Portman, Healing Manor, at 33gs., while Lord Algernon Percy, Guy's Cliffe, Warwick, gave 30gs. for Foston Gem II., by Dowsby Waterloo Oliver II. (C.H.B. 72,372). Foston Sybil, by Foston Duke (2204) went to Captain Portman at the last-mentioned figure. Mr. John Evens, Burton, Lincoln, gave most money for two-year-old heifers, getting Foston Blossom, a handsome daughter of Belchford Boy 1st (3735), and a cow by Foston Duke (2204) knocked down to him at 35gs.

1909. There were several good private auction sales in 1909, and the average throughout were most satisfactory. Mr. George Freir held another sale on September 30th, when a choice selection were sold by Messrs. Dickinson, Riggall & Davy, at Deeping St. Nicholas, near Spalding. The thirty-five Lincoln Reds averaged £25 4s. A heifer calf by Buscot Rupert was sold to Mr. A. P. Brandt, Bletchingley Castle, Surrey, for 32gs., and Mr. Sharman paid 35½gs. for the cow Hurn Wallflower 2nd, by Surfleet Beau (2667). Mr. Brandt bought the two best yearling heifers at 28gs. and 27gs. Mr. A. P. Brandt also gave 28gs. for Deeping Stenigot, by Stenigot Duchess Gwynne (5633), and 27gs. for Deeping Daisy 3rd, by Bigby Genial (5155).

A dispersal sale of Lincolnshire Red Shorthorns, the property of Colonel C. A. Swan, was conducted by Mr. J. E. Walter, at Aswardby Hall, near Spilsby, on Thursday, October 7th. The twenty cows and calves averaged £33 19s. 10d. Captain C. L. Prior bought a handsome cow by Scampton Angler (2327) at 31gs., and Mr. E. N. Crowder purchased another very useful cow by Bonby Wolseley XVII. (3233) at 33gs. Mr. Crowder, who was a most extensive purchaser, gave 32gs. for a sweet two-year-old heifer by Hainton Favourite (4403) and 30gs. for another.

The Summary of the sale is as follows :—

	Average. £ s. d.	Total. £ s. d.
32 Cows, heifers and calves.. ..	30 2 9 ..	964 8 6
1 Bull	36 15 0 ..	36 15 0
33	£30 7 1	£1,001 3 6

H

The dispersal of the herd belonging to Messrs. H. M. & R. T. Proctor was conducted by Messrs. Benjamin Simons & Co., at Wykeham House, near Spalding, on October 14th. The two-year-old heifers were a very good lot, with good hair and colour, and matching like peas. The twenty-three cows and calves averaged £28 17s. 3d. Mr. F. Scorer took the pick in Wykeham Fillpail 3rd, whose dam, Wykeham Fillpail, Mr. H. M. Proctor declared gave nine gallons of milk a day in her prime. This old cow also went to Nettleham as did her daughter, Wykeham Fillpail IV. by Red Monarch (C.H.B. 77,605), she being knocked down at 38gs. Mr. M. Staniland secured the two-year-old heifer Wykeham Maiden 2nd, by Benniworth 7th (1458), at 35gs., several other lots falling to him ; while Mr. Joseph Bowser was a most extensive purchaser, and Colonel Morgan and Messrs. A. H. Clark, W. J. Atkinson, G. Freir, F. Mowbray, J. Martin, J. G. Shepherd, and E. W. Farrow were also good customers.

The Summary of the Sale is as follows :—

		Average.			Total.		
		£	s.	d.	£	s.	d.
75 Cows and heifers ..	.. ..	22	1	0	1,654	0	6
2 Bulls	.. ..	17	1	3	34	2	3
77		621 18 5			£1,688	2	9

Messrs. Mason, Sons & Kelk, and Messrs. Dickinson, Riggall & Davy conducted the dispersal sale of the herd of Lincolnshire Red Shorthorns, the property of Messrs. J. T. and the late A. W. Needham, at Huttoft Grange, near Sutton-on-Sea, on October 26th, the sale being necessary on account of the death of Mr. R. W. Needham. The first two cows were sold to Mr. C. Bett, Benniworth, and Mr. Johnson, at 19gs. and 19½gs., both being by Huttoft Nimrod (2547), and then Mr. G. E. Collins, Caistor, secured one by the same sire, a famous breeder and a great milker, cheaply at 22gs. Mr. Johnson, Rugby, also gave 21gs. for a similarly-bred cow. Mr. E. N. Crowder, Horncastle, paid 30gs. for a cow by Huttoft Nimrod, and he bought others at 22gs. and 21gs. The old bull, Huttoft Ben (3938), by Lord Bobs (2965), was knocked down to Mr. C. Dickinson, at 24gs., and his sixteen months old son from Mr. Crowder's 30-guinea purchase was sold to Mr. J. W. Davy, at 26½gs.

A Summary of the sale is appended :—

					Average. £ s. d.			Total. £ s. d.	
25	Cows and calves	..	..	..	26 16 2	..	670	4	3
12	Two-year heifers	..	..	..	19 17 8	..	238	12	3
16	Yearlings	..	..	..	12 2 9	..	194	5	0
3	Bulls	..	..	..	22 15 0	..	68	5	0
56					£20 18 3		£1,171	6	6

The dispersal of the celebrated herd of Lincolnshire Red Shorthorns, the property of Mr. Robert Chatterton, was conducted by Messrs. Dickinson, Riggall & Davy, at Stenigot, near Louth, in Lincolnshire, on Thursday, October 28th, when the eighty-eight head obtained an average of £37 3s. 8d., a very remunerative one when it is remembered that it was a dispersal sale, and not one of a choice selection. No sensational figures were obtained, but prices ruled satisfactory throughout, showing that the cattle can still be bought at reasonable sums. The cows with bull calves were a wonderfully good lot, selling at the good average of £50 4s. 9d. Mr. Achurch paid 74gs. for the fine cow Stenigot Buttercup 10th, by Sirdar (1676), and her calf by Keddington Comet (3443), and Primrose 25th, by Red Chief (2611) went to Mr. Wingfield, at 42gs., while Mr. A. P. Brandt gave 40gs. for a bonny calf by Keddington Comet (3443) out of Stenigot Duchess 3rd. Lord Cole gave 37gs. for a handsome cow to go to South Africa. Good prices, too, were made for cows with heifer calves. The bulls should have made more money, and the Royal winner, Stenigot Duke, by Stenigot Duchess Gwynne (5633), was certainly cheap at 80gs., the figure at which he was knocked down to Mr. F. Mowbray. The highest prices and averages are appended :—

COWS AND BULL CALVES.

	Gs.
Waltam's Beauty—J. Byron	30
Her b. c.—Mr. Blaydes..	35
Stenigot Milker—Lord Cole	37
Her b. c.—Mr. Birkbeck	34
Stenigot Buttercup 10th and her b. c.—Mr. Achurch	74
Her b. c.—F. Martin	37
Stenigot Milkmaid—Mr. Neave	32½
Her b. c.—J. R. Dennis	38
Stenigot Primrose 25th—Mr. Winkfield	42
B. c. by Keddington Comet—A. P. Brandt	40
B. c. by Keddington Comet—Mr. Doncaster..	33
Kirmington Briar 11th and her b. c.—Col. Morgan	53

Gns.

Stenigot Ruby 14th—Mr. Campion 33
Stenigot Ashby 3rd—Mr. Neave 31
Stenigot Daisy 15th—Mr. Ranby 38
Brandon Nonpareil Ruby—J. G. Williams 31

COWS WITH HEIFER CALVES.

Stenigot Daisy 20th—J. Byron 33
Brandon Ruby Nonpareil—G. E. Sandars 41
H. c. by Gwynne Cowslip—G. Laughton 30

TWO-YEAR-OLD HEIFERS.

Stenigot Daisy 17th—F. B. Wilkinson 42
Stenigot Crocus—Mr. Crowder 32
Stenigot Red Rose 37th—A. P. Brandt 52
Kirmington Rose 21st—F. B. Wilkinson 38
Stenigot Choice 15th—Col. Morgan 33

YEARLING HEIFERS.

Stenigot Croft Ruby—T. Turner 30½
Alpine Bella (C.H.B.)—E. H. Cartwright 32

BULLS.

Stenigot Duke—F. Mowbray 80
Stenigot Comet—Mr. Haywood 62
Maori Duke—Mr. Fletcher 30
Stenigot Gwynne Comet—Mr. James 35
Stenigot Bloom Cowslip—T. Harrison 30
Stenigot Gwynne Cowslip—E. Smith 40
Saltfleet Red Knight—H. Gaunt 36
Hallington Neptune—F. B. Wilkinson 43

SUMMARY.

		Average.				Total.		
		£	s.	d.		£	s.	d.
23	Cows with bull calves	50	4	9	..	1,155	10	6
22	Cows with heifer calves	40	8	0	..	888	16	6
14	Two-year-old heifers	30	8	3	..	425	15	6
18	Yearling heifers	19	3	10	..	345	9	0
94	Cows and calves	36	6	0	..	3,412	10	0
11	Bulls	41	10	5	..	456	15	0
88	Head	£37	3	8	..	£3,272	6	6

1910.   Messrs. B. Simons & Co. conducted a sale of Lincolnshire Red Shorthorns, the property of Messrs. J. H. Wright & Charles Parker, at New Leake, near Boston, on April 5th, *1910*. Mr. W. Caudwell, Holbeach, gave 20gs. for a cow in-calf to Partney Victor (6246), and another fell to Mr. Joseph Bowser, Frithville, at 21gs., at which figure Mr. F. Spencer, Folkingham, was a buyer, while Mr. J. T. Harrison,

Midville, paid 20gs. for another. Mr. John Searby, Croft, and Mr. Ben Rowlands, Wainfleet, were also buyers of cows, Two-year-old heifers made up to 16½gs.

The Lincoln Reds, a draft from the herd of Mr. H. Abra-ham, Walesby, which were offered by Messrs. Dickinson, Riggall & Davy, on October 4th, were a very useful selection, and with several prominent breeders present some good prices were obtained, the general average for 25 head being the very satisfactory one of £26 19s. 5d. The first lot to come into the ring was Walesby Stella, a lengthy, wide, deep cow by Benni-worth 21st (2742) (a bull bred by the late Mr. T. Bett) out of a Dunsby Sentinel (1535) dam, one of a rare milking strain. Both Mr. G. Marris (the well-known breeder from Kirmington) and Lord Heneage wanted her, but she was eventually purchased for the Hainton Hall herd, at 40gs. Lord Heneage made two other judicious purchases, paying 30gs. for Walesby Milkmaid, by Scampton Cypress (3048), a wealthy, good cow with a capital bag, and reputed to give six gallons of milk a day ; and 29gs. for Walesby Dairymaid, a level, lengthy cow by Dunsby Sentinel (1535). Mr. G. Marris also secured a good cow in Walesby Motley, by Benniworth 21st (2742), at 27gs., and another of a fine milking strain, Walesby Favourite, by the same sire, was knocked down to him at 25½gs. The six handsome two-year-old heifers sold well, Mr. Marris taking the pick in Walesby Primrose, by Benniworth 21st (2742) from Healing Miss Wainfleet, a well-bred one that cost him 31gs. ; while the plum of the half-dozen very promising yearlings was knocked down to Mr. E. Sharpley, Kelstern, at 20gs. This was Walesby Red Rose, by Strubby Red Coat 1st (5647), out of Walesby Stella. Mr. W. J. Need, Winthorpe Hall, Newark, was a good customer ; and Mr. W. Chatterton, Castlethorpe, bought several lots. The stock bull Strubby Red Coat 1st (5647), by Under Porter (3126) was sold to Mr. Topliss, Market Rasen, at 30gs.

The Summary of the sale is as follows :—

				Average.				Total.		
				£	s.	d.		£	s.	d.
12	Cows and heifers..	..	..	31	11	3	..	378	15	9
6	Two-year-heifers ..	..	..	25	16	3	..	154	17	6
6	Yearling heifers ..	..	..	18	4	0	..	109	4	0
1	Bull	..	..	31	10	0	..	31	10	0
25				£26	19	5		£674	7	3

There has been no event that has demonstrated the increasing popularity of the Lincoln Reds outside their native county so much as the sale which took place at Pendley Manor, Tring, on Tuesday, October 18th, 1910, when a choice selection from the herd belonging to Mr. J. G. Williams, J.P., was offered by Messrs. Dickinson, Riggall & Davy, and Messrs. W. Brown & Co. Mr. Williams' herd was founded in 1898 by purchases from the herd of Mr. John Evens, Burton, and nothing but the very best blood has been added since, the judicious purchases from Croft, Benniworth, Keddington and Stenigot, accounting for the truly exceptional success which the Pendley Manor herd has had at the Royal and Lincolnshire Shows. There was a good attendance of Lincolnshire breeders and many of the best animals went back to the county, but it is gratifying to see that purchases were also made to go into South and West-country herds, where the true worth of this, the premier dual-purpose cattle, is bound to make itself known.

The sale was in every way a great success and must rank with those held at Keddington and Stenigot as landmarks in the history of the breed, a very satisfactory recognition of Mr. Williams' pluck and enterprise and the skill and management of his agent, Mr. H. W. Bishop.

The first cow to make a good price was Benniworth Bloom, by Saltfleet Actor (1664), who was knocked down to Mr. John Evens, Burton, at 40gs., while Mr. G. Palmer, Lacock, Wilts. secured a grand cow of the milking type in Enderby Lass IV. by Lord Chancellor (1606). Mr. T. H. B. Freshney takes a most promising bull calf by Keddington Comet (3443) out of Benniworth Violet II. to South Somercotes, at 43gs., and another capital youngster, Pendley Knight, by Stenigot Cornish Knight (C.H.B. 104,059) from Stenigot Gwynne XVI. fell to Mr. J. W. Measures, Bourne, at 46gs. Colonel Morgan, who is getting together a choice herd of Lincoln Reds, at South Leverton Priory, Notts., took Dunsby Nonpareil, by Weston Nonpareil King (2068), of great dairy character, at 42gs., and her bonny bull calf by Sausthorpe Red XII. (4552) became the property of Mr. P. J. Hensman, Fulletby, Horncastle, at 30gs., Mr. F. Scorer, Nettleham, also paying 35gs. for a little beauty by Keddington Comet (3443) from Pendley Rose.

But the top price of the sale was made by the bull calf by Keddington Comet (3443) out of Keddington Skipworth V., by Benniworth IV. (629), who after a spirited contest between Mr. E. H. Cartwright, Mr. J. W. Measures, and Mr. J. Evens, fell to the former at 102gs. The dam of this calf, who must have a great future before him, was female champion at the Lincolnshire Show, in *1909*, and reserve to her herd-mate, Pendley Starlight, this year. Pendley Pearl, by Saltfleet Echo (3038), first at the Royal both this and last year, went to Mr. Palmer, at 50gs. and must do much to enhance the reputation of the Lincoln Reds in Wiltshire, and Colonel Morgan made another judicious purchase in Keddington Chance, for whom he paid 40gs. The average of £45 7s. 1d. for 23 cows and calves is only such as they deserved ; but the nine sweet two-year-old heifers might have made a little more money. There was a ready sale for the handsome selection of yearling heifers. the first of which Pendley Pearl II., by Bonby Excursionist IV. (5161), fell to Mr. Palmer, at 40gs., but the pick of the bunch, Pendley Lassie, by Nonpareil Bonus (6216) out of Keddington Lassie II., by Keddington Baron (4881), second at Peterborough and Spalding, was most admired. Mr. F. B. Wilkinson, Mr. Ekins Ashley, Godmanchester, and Mr. P. J. Hensman, were all in the bidding, but Mr. E. H. Cartwright had the last word at 72gs.

The average of £30 2s. 3d. for the yearlings was most thoroughly deserved, while the general average of £38 1s. 3d. for 44 head was most encouraging to breeders of what has been rightly termed the ideal tenant-farmers' cattle. Among the buyers were Mr. Vicars, Leamington, Messrs. J. N. Robinson & Son, Anderby, Huttoft, Mr. G. Freir, Deeping St. Nicholas, Mr. G. Marris, Kirmington House, Brocklesby, Mr. J. K. Foster, M.P., Mr. A. P. Brandt, Bletchingley Castle, Surrey, Mr. E. Crowder, Aswardby Hall, Mr. Chandos de Paravacini, Grantham, and Mr. G. Freir, Deeping St. Nicholas.

The highest prices and averages are appended :—

COWS AND CALVES.

Purchaser.	Gs.
Benniworth Bloom—J. Evens, Burton	40
Enderby Lass IV.—G. Palmer, Lacock, Wilts.	37
Her b. c. by Keddington Comet—T. H. B. Freshney, South Somercotes	43
B. c. by Pendley Knight—J. W. Measures, Bourne	46
B. c. by Keddington Comet—F. Scorer, Nettleham..	35
Dunsby Nonpareil—Col. Morgan, S. Leverton Priory, Notts. ..	42
Her b. c. by Sausthorpe Red XII.—P. J. Hensman, Fulletby ..	30
B. c. by Keddington Comet—E. H. Cartwright, Keddington Grange	102
Pendley Pearl—G. Palmer	50
Pendley Dairymaid—C. de Paravacini, Grantham	23
Keddington Chance—Colonel Morgan	40
Pendley Village Maid—J. Evens	35

BULL.

Grange Prince—E. H. Cartwright	50

TWO-YEAR-OLD HEIFERS.   Gns.

Pendley Wanton III.—Mr. James, Herts. 42
Pendley Rosebud III.—Mr. James 38

YEARLING HEIFERS.

Pendley Pearl II.—G. Palmer.. 40
Pendley Dairymaid II.—F. B. Wilkinson 37
Pendley Lassie—E. H. Cartwright 72

SUMMARY.

		Average.				Total.		
		£	s.	d.		£	s.	d.
23	Cows and calves	45	7	1	..	1,043	3	6
9	Two-year-old heifers	27	10	8	..	247	16	0
11	Yearling heifers	30	2	3	..	331	5	6
1	Bull	52	10	0	..	52	10	0
44		£38	1	3	.	£1,674	15	0

Messrs. Dickinson, Riggall & Davy, in conjunction with Messrs. Royce, conducted another sale of Lincoln Red short-horns, at Londonthorpe, near Grantham, on October 13th. The cattle had not been got up for sale in any way, and were shown in poor condition. Some of the calves were under two months old. 36 cows and heifers with calves averaged £24 5s. 6d., in-calf heifers averaged £17 10s. 3d., 20 one-and-a-half-year-old heifers averaged £14 0s. 10d. The yearling heifers were very low in condition, and made up to £10 10s. There was a large company present, and most of the animals went into Rutland, Nottinghamshire and Leicestershire.

A summary of the chief private sales is here appended :—

PRIVATE AUCTION SALES.

1900. Herd.	No. sold.	Highest price. Gs.	Average. £	s.	d.	Total. £	s.	d.
Messrs. S. & J. Crawley ..	61 ..	42 ..	25	19	7 ..	1,584	19	6
Mr. C. H. Stafford	41 ..	57 ..	33	5	10 ..	1,364	18	6
1901.								
Mr. W. J. Atkinson	77 ..	40 ..	21	0	0 ..	1,604	18	6
Messrs. R. & R. Chatterton..	89 ..	110 ..	35	10	10 ..	3,163	2	6
1904.								
Mr. Everett King ..	104 ..	45 ..	18	14	6 ..	1,947	15	0
Mr. W. J. Atkinson..	65 ..	120 ..	27	4	0 ..	1,768	4	0
Mr. G. Frier	70 ..	50 ..	23	9	11 ..	1,644	16	6
1905.								
Messrs. S. E. Dean & Sons ..	81 ..	40 ..	27	11	5 ..	2,222	6	6
Mr. J. Marriott ..	61 ..	36 ..	16	12	6 ..	955	8	0
Mr. A. Smith.. ..	83 ..	46 ..	20	8	8 ..	1,696	5	6

1906. Herd.	No. sold.	Highest price. Gs.	Average. £	s.	d.	Total. £	s.	d.
Mr. T. B. Freshney ..	.. 56	.. 106	.. 21	0	8	.. 1,198	1	0
Mr. H. C. Tinsley ..	.. 86	.. 36	.. 16	13	4	.. 1,433	10	3
Mr. T. Bett	.. 57	.. 50	.. 25	9	2	.. 1,451	12	6
Mr. E. H. Cartwright	.. 65	.. 104	.. 40	9	0	.. 2,629	9	3
1907.								
Mr. G. Freir	.. 48	.. 50	.. 26	19	0	.. 1,293	12	0
Mr. W. A. Ewbank ..	.. 72	.. 36	.. 23	14	8	.. 1,708	18	3
Mr. J. Crawley ..	.. 42	.. 91	.. 23	11	9	.. 990	13	6
Mr. G. Laughton ..	.. 38	.. 46	.. 23	4	2	.. 882	0	0
Messrs. J. W. Farrow & Sons	71	.. 42	.. 30	13	7	.. 2,178	9	9
Mr. W. J. Atkinson ..	.. 63	.. 52	.. 34	2	8	.. 2,150	8	0
Mr. J. Langham ..	.. 65	.. 52	.. 23	3	7	.. 1,506	15	0
Mr. R. Chatterton ..	.. 66	.. 99	.. 43	10	4	.. 2,872	6	6
1908.								
Mr. J. Searby ..	.. 60	.. 56	.. 27	7	0	.. 1,641	3	0
Mr. G. Marris ..	.. 34	.. 50	.. 23	15	3	.. 760	9	0
Mr. T. Gaunt ..	.. 45	.. 31	.. 24	4	0	.. 1,089	7	6
Mr. J. T. Cox ..	.. 46	.. 38	.. 21	4	0	.. 975	4	0
1909.								
Mr. G. Freir	.. 43	.. 35½	.. 24	5	5	.. 1,043	14	0
Messrs. J. T. & A. W. Need- ham ..	.. 56	.. 30	.. 20	18	3	.. 1,171	6	6
Mr. R. Chatterton ..	.. 88	.. 80	.. 37	3	8	.. 3,272	6	6
Messrs. H. M. & R. T. Proctor	77	.. 38	.. 21	18	5	.. 1,688	2	7
Col. C. A. Swan ..	.. 33	.. 35	.. 30	7	1	.. 1,001	3	6
1910.								
Mr. H. Abraham ..	.. 25	.. 40	.. 26	19	5	.. 674	7	3
Mr. J. G. Williams ..	.. 44	.. 102	.. 38	1	3	.. 1,674	15	0

SOME OF THE LEADING HERDS.

SOME OF THE LEADING HERDS.

Of the breeders of Lincolnshire Red Shorthorns, both before and since they became a registered breed, none have done more to maintain the character of the cattle and to bring them to a greater state of perfection than have the Messrs. Chatterton, of Stenigot and Hallington, Mr. E. H. Cartwright, of Keddington, the late Mr. T. B. Freshney, of South Somercotes, Mr. G. E. Sandars, of Scampton, and Mr. John Evens, of Burton ; while for many years it was left to the latter alone to demonstrate the magnificent dairy properties of the breed, and till quite recent years he has fought the battle of the milk pail against all comers single-handed. Now Messrs. F. & C. E. Scorer (father and son) have come into the field, and are helping on the good work, to their own and the breed's advantage. In my first article I showed how the cattle were descended from the purest C.H.B. strains, and that judicious introductions of the blood had been made ever since, resulting in more Shorthorn character, but without losing those points which have made the cattle of Lincolnshire so justly famous, their hardiness, thriftiness, economy in feeding, wealth of lean flesh, and exceptional milking powers. I have already referred to the influence, in the building up of the breed, of Mr. Turnell, of Reasby, Mr. Coulam, of Withern, Mr. Baumber, of Somersby, Mr. Oliver, of Eresby, Messrs. Burtt, of Welbourne, Mr. Cartwright, of Tathwell, and the Messrs. Chatterton ; but no history of the Lincoln Reds would be complete without a reference to those who followed in the footsteps of these pioneers, and are now in the forefront of the breeders of this ideal tenant-farmers' cattle. And so, after some careful thought, and the study of the innumerable facts and figures which go to make up the brief history of the breed which it has been my pleasure to write, I have come to the conclusion that none deserve more credit among modern breeders for the present position of the Lincoln Reds than do those whose names appear in the first few lines of this chapter.

Mr. E. H. Cartwright's herd came into the possession of the owner in the spring of *1881*, part of it being obtained by the purchase of 22 cows and heifers at the sale of Mr. H. Peacock, of Ulceby, near Alford, who had succeeded to his father's herd, which had been in existence many years ; and he also bought 12 females at the sale of the late Mr. Adam Eve, of Haugham. Since then purchases have been made from Mr.

C. Robinson, Bilsby, Mr. W. Chatterton, Hallington, Mr. J. Chapman, Louth, Mr. W. Enderby, Louth, Mr. C. Skipworth, Tathwell, Mr. J. Mason, Calceby, and Mr. T. Swingler, Langham, near Oakham (from whom he bought three cows of the old Turnell blood), and other famous breeders from year to year. Previous to 1895, Mr. Cartwright used such sires as Windsor Benedict (C.H.B. 40,933), bred by Mr. T. Willis, Carperly Manor, Bedale, Yorks. ; Ludford (172) bred by Mr. G. E. Sandars ; Worlaby Lad (289) bred by Mr. J. Mason ; Woodhall (286) and Astronomer (21) bred at Keddington ; Admiral (3) bred by Mr. G. Grifin ; and Orangeman (190) bred by Mr. J. Drewery. The celebrated sire Bigby (319), who was bred by Mr. Cartwright in 1892, and had been used by Mr. George Walker, of Bigby, with great success, was used in 1896, and again in 1897 (in which year additions to the herd were made through the purchase of some females from Mr. W. Chatterton's herd), and he continued to use him till 1899, when Grandad (156) was also used. Two more famous bulls, Benniworth 4th (629) bred by Mr. T. Bett, Benniworth, and Conisholme Boy (347), bred by Mr. J. B. Hill, Smethwick Hall, Congleton, Cheshire, were used in 1900 ; and the weighty Vanguard (2691), together with the first named, in 1901. Vanguard was a son of Benniworth 4th (629) and was bred by Mr. G. A. Oliver, Hallington ; and both father and son, were used in 1902, and again in 1903, with other bulls. The sires used in 1904 were Brandon Lord Chief Justice (2425) of Mr. J. Langham's breeding ; Horkstow Hero (3422) bred by Mr. W. B. Swallow, Wootton Lawn, Ulceby ; the mighty Saltfleet Bonus (3582), who once defeated Mr. Philo Mills' famous bull King Christian of Denmark for the championship of the Lincolnshire Show, and who was bred by Mr. T. B. Freshney ; and Stenigot Bloom Boy (3611), who was bred in 1902, by Messrs. R. & R. Chatterton. Saltfleet Bonus and Keddington Bloom Boy were also used in 1905 ; and in 1906 Keddington Baron (4881) and Keddington Comet (3443), both home-bred ones, were the two bulls used. A perusal of the previous articles will show what the Keddington blood has done in the show and sale rings, stamping it as being both as fashionable and as remunerative as any in the Herd Book. By no means the least important of the animals bred by Mr. E. H. Cartwright was Keddington Ruby (1243), whose sons, bred by Mr. G. E. Sandars, Scampton, have averaged more money at the L.R.S.A. sales at Lincoln April Fair during the past few years, than those of any other bull. He was by Bigby (319), out of Keddington Skipworth 2nd, by Commodore (81), dam Worlaby Skipworth, by Ludford (172), grand-dam Haugham Skipworth, by Windsor Benedict (C.H.B. 40,933); Bigby being by Orangeman (190) from that celebrated cow, Nonpareil, by Hyllus (149). Nonpareil Bonus (6216), by Knockaabout (4899), bred at Keddington, was the stud bull in 1909.

Judicious purchases of Keddington cattle had no little to do with the show-yard successes of Mr. J. G. Williams Pendley Manor, Tring. The highly important sales at Keddington have been referred to elsewhere.

The wonderful success of the Stenigot cattle in the show ring and the high prices made of them at the occasional sales held at their home between Louth and Lincoln, is sufficient proof of their merits. Such cattle as was formerly bred by Messrs. R. & R. Chatterton and afterwards, till the sale in 1909, by Mr. Robert Chatterton, would be a credit to any breed. The herd came into the possession of the two brothers in 1885, and was obtained from the herd of their uncles, Messrs. Richard & William Chatterton, which had been in existence most of a century. The foundation of the present herd may be said to have been laid with the purchase of a famous cow from Mr. Coulam, of Withern, who dropped a red heifer calf called Alcama. This latter was sent in 1892 to the Marquis of Exeter's famous bull Cambridge Duke 5th (C.H.B. 30,644), who combined the Duchess and Red Rose blood, the result of the union being the bull Hercules (144), who was used in the herd for nine seasons and proved a most famous sire. His son Hyllus (149)—both father and son having been bred by the late Mr. William Chatterton, at Hallington—was also used in the herd till the sale in 1885, at which sale Messrs. R. & R. Chatterton purchased 109 of the females. These were mated with the best Hallington and C.H.B. Red Shorthorns, and at the time of the formation of the Lincolnshire Red Shorthorn Association in 1895 there were in the herd two cows by Hyllus (149), six by Candidate (56), four by Kinsman 21st (C.H.B. 59,206), ten by Crœsus (85), who was by Hyllus from a most noted Hallington cow Riches, 18 by Comet (79), who was by Eclipse (111), by Windsor Benedict (C.H.B. 40,933) out of a daughter of Riches, and one by Eclipse; there were also four heifers by Eclipse, four by Commodore (81), one by Hallington (135), three by Admiral (3), seven by County Member (83) a son of Crœsus (85), and three by Ballywalter (23), who was by Lancelot (C.H.B. 61,094), the herd then consisting of 41 cows calved previous to 1893, and 22 heifers calved in 1893 and

1894.   The herd in 1906 numbered 92 cows and heifers and 37 calves, calved that year.   Space will not permit mention of the many famous animals bred and used at Stenigot, and so I will but call to mind the names of a few whose doings have been already recorded in previous pages.   Conisholme Boy (347) was used in 1898, and also his son Sirdar (1676), who has written his name large in Lincoln Red history, and whose breeder was Mr. J. B. Hill, Smethwick Hall, Congleton, Cheshire ; Volunteer (1715), by Red Prince (211) from a Crœsus (85) dam, who afterwards went to Argentine ; Wolseley (1436) who was by Volunteer (C.H.B. 63,501) and was bred by the Earl of Yarborough ; Wrangler (C.H.B. 71,901) ; Red Chief (2611), by Sirdar from Stenigot Red Rose 3rd, bred by the Messrs. Chatterton ; Head Porter (2909) who was bred at Hallington, and was by Nonsuch (1292) from a Red Prince (211) cow ; Stenigot Beau 2nd (4107), by Wolseley ; Ashby Red 2nd, by Ashby Wrangler No. 1 (3184) from Stenigot Nonsuch, and bred by Mr. E. G. Wattam, West Ashby, Horncastle ; Cornish Knight (C.H.B. 78,641) ; and the home-bred Stenigot Gwynne Chief (4996), by Red Chief (2611) out of Stenigot Gwynne 4th.   But to write the history of the Lincoln Reds is to write the history of the Stenigot herd, for the one is bound up in the other.   The late Mr. John Thornton used to say that it was not the getting together of a herd that did the good to breeding, but the dispersal of it, and the two great sales of 1907 and 1909, though meaning the breaking up of a famous herd, will have a great influence on Lincoln Red breeding.

The history of Mr. William Chatterton's herd at Hallington, is practically the same as that of the Stenigot herd, it having been founded from purchases from the late Mr. W. Chatterton's herd in 1885, and that of the late Mr. R. G. Chatterton's at Tathwell, which herds by the way, had for years been using bulls bred by Mr. Richard Dudding, of Panton, Mr. Sharpley, of Calcethorpe, Mr. Coulam, of Withern, Mr. Walesby, of Ranby, Mr. Cartwright, of Haugham, the Marquis of Exeter, and others.   Mr. W. Chatterton's bulls have always been in great demand by the leading breeders, and the important part they have played in building up the Lincoln Red breed of to-day must have been noticed by those who have been reading my little work.   The celebrated Eclipse (111), by Windsor Benedict (C.H.B. 40,933) out of Luna, by Hercules (144), who was by the Marquis of Exeter's noted bull Cambridge Duke 5th (C.H.B. 30,644) from Alcama, the daughter of the famous cow, bought of Mr. Coulam, of Withern, was both bred and used at Hallington, and stands as one of the landmarks in Lincoln Red history ; and Mr. W. Chatterton also bred King Crœsus, by Kinsman 21st (C.H.B. 59,206) out of the noted cow, Riches ; Rufus (225) another son of Eclipse and Red Prince, by Commodore (81) from an Eclipse (111) dam. Bigby (319) was used at Hallington in 1896 ; and Nonsuch (1292), who was by Volunteer (C.H.B. 63,507) from that famous cow Nonpareil, by Hyllus (149) and was bred by the Earl of Yarborough, in 1897 and 1898, and again in 1899, together with Conisholme Boy (347), and Negligence (1626), a son of the last named, and bred by the breeder of the sire. The chief sire in 1900 was Nono (2274), by Nonsuch ; and some of Mr. T. B. Freshney's blood was introduced the next year through Saltfleet Ascent (1671), a son of Bigby (319), and Saltfleet Combatant (2319), by Grandad (1561). Then in 1902, both Nero (2991) who was by Red Monarch (C.H.B. 77,605) from a Hyllus (149) cow, and was bred at Weston St. Mary, by Mr. W. J. Atkinson ; and Walmsgate Mate 2nd (1722) of Messrs. Simons' breeding were used ; and the former was in service for the next three years, making his mark in the herd, as did Horkstow Hero (3422) who was of Mr. W. B. Swallow's breeding. Bonby Excursionist 26th (5846) of Messrs. T. & W. Dickinson's breeding, and Guardian (4397), a Keddington-bred bull were in service in the herd in 1909.

The late Mr. T. B. Freshney's Ruby family is one of the best known of the Lincoln Reds, and it has made a name for itself both in the show-yard and the sale ring. The South Somercotes herd was founded in 1885 from purchases of cattle belonging to Mr. W. L. Mason, of Keddington, Mr. Richardson, of Baumber, and Mr. Stratton, of Manningford Bruce, Wilts., a small but a select collection of cattle ; and the sires used at the time of the formation of the breed Society were Saltfleet South Ormsby (229) bred by Messrs. Mundy & Ward, Saltfleet Eclipse (227), by Eclipse (111) bred by Mr. W. Chatterton, and Saltfleet Keddington (228), by Admiral (2), bred by Mr. J. Reeson, Keddington. Later on Mr. Freshney used Saltfleet Sappy (502), Saltfleet Saturn (503), Saltfleet Mars (1342), Saltfleet Prince Ruby (942) and Saltfleet Sentinel (945) ; and in 1898 such bulls as Bigby (319), Grandad (1561), Saltfleet Electric (1338), and Shooting Star (1674) were leaving their mark in the herd. Another good bull used in 1899 was Saltfleet Ascent (1671), and he and Grandad were responsible for the 1901 crop of calves. Then came that impressive bull

I

Benniworth 4th (629) the next year ; but the bull of all others that will be associated with the name of Mr. T. B. Freshney, is Saltfleet Bonus (3582) a very massive, heavy-fleshed bull by Red Monarch (C.H.B. 77,605) from a cow by Lord Knightley (170), which latter was by Cambridge Duke 29th (C.H.B. 60,440) and was bred by Mr. Henry Sharpley, at Limber. Saltfleet Bonus (3582) was bred by Mr. F. Riggall, at Well, Alford, was several times winner at the Royal and Lincoln-shire Shows, and in 1905 was awarded the championship of the Lincolnshire Show, beating Mr. Philo Mills' famous bull King Christian of Denmark, who won many championships all over the country, and was sold at the dispersal sale of the Ruddington herd to Mr. A. W. Hickling, Adbolton, Notting-hamshire, for 900gs. Saltfleet Bonus was used both in 1903 and 1904 ; and in the latter year some of the best matrons of the herd were served by Imperial Favourite (C.H.B. 86,233), the handsome young red bull for whom Messrs. S. E. Dean & Sons paid 700gs. as a calf at the late Mr. W. S. Marr's sale in Scotland, and afterwards sold in Argentina for the equivalent of 1,000gs. Saltfleet Bonus was used again in 1905 and 1906, and in the latter year his sons Somercotes Bonus (4577) and Saltfleet Dragoon (4547). Mr. T. H. B. Freshney is now carry-ing on his father's herd with great success, and the bulls he used in the herd in 1909 were Grange Prince (4843) bred at Keddington, Saltfleet Imperialist (4549) a home-bred one by Imperial Favourite (C.H.B. 86,233), and Saltfleet Red Knight (5604), by Keddington Comet (3443).

Mr. John Evens' now world-famous dairy herd has descended from father to son for generations, and he took it over himself in 1875. His wonderful successes in the show-yards and at the milking trials and butter tests in England and Ireland have been told elsewhere, and it will be sufficient to indicate a few of the strains of blood that have been intro-duced into the herd through the sires. Mr. Evens believes that the sire is more than half the herd ; and the bulls he selects for service have not only themselves to be from great milking dams, but the sires of these bulls also. A good bull used about the time of the formation of the Association's herd register was Ramsden (203), who was bred by Sir John Ramsden, Bart., at Cotes, near Gainsborough ; and Fox (122), by bred Mr. E. Burtt ; Khartoum 2nd (153) bred by Mr. H. Burtt ; and Burton Butterfly (C.H.B. 63,741) bred at Burton, were also used at that period. Professor (200) a home-bred bull by Fox, also proved a successful sire ; also Knight of Chewton

(C.H.B. 68,867) ; Butter Boy (1477), by Pretender (C.H.B. {64,544) ; Burton Profit (1108) ; and Burton Royal Profit (1805), by Rambler (C.H.B. 69,342) out of a cow by Beauty Bull (1770), a son of Mr. Evens' first winning cow, Burton Beauty, who in 34 months gave 3,673 gallons of milk, and in 1887 won first prize in the milking trials at the London Dairy Show and carried off the Lord Mayor's Champion Cup. Old Profit the dam of Burton Royal Profit (1805), was a Royal winner, produced six calves in eight years, and yielded 8,440 gallons of milk. Red Rover (C.H.B. 77,618), who cost Mr. Evens £100 at the London Dairy Show in 1899, and was a first prize-winner the only two occasions on which he was shown, was the sire of innumerable dairy winners, and was himself from a dairy cow that won 23 prizes, including milking trials. Other good bulls used were Burton Rex (2131), who weighed 21¼cwt., and whose dam, Coa Fox, yielded over 5,000 gallons of milk after her five calves ; Scampton Expansion (4093) a Royal winner, and from the best dairy cow in Mr. G. E. Sandars' herd, for whom Mr. Evens paid 130gs., and afterwards sold at a high price to go to Chili ; and Burton Challenger (3265), for whom Mr. Evens refused £150 when 10 months old. This bull's dam, Burton Margaret, was the biggest and best dairy cow Mr. Evens ever owned, and she holds the record for milk at the Tring Milking Trials, the largest in England, viz., 7½ gallons in 24 hours. Burton Challenger's sire was Bracebridge Ormsby 2nd (2772), who was bred by Mr. F. Scorer the owner of another dairy herd that is rapidly coming into the front rank. Burton Carl (3777) and Morello (4481), used in 1905 and 1906, were both by Burton Rex (2131), and the most recent stud bulls have been Burton Challenger 2nd (4723), by Burton Challenger (3265) from Marjorie. Mr. Profit (4926) by the same sire out of Profit, and Burton Prime 2nd (4287), by Burton Rex (2131) out of Primrose 2nd, each of the dams being great winners at Dairy Trials and Butter Tests. Profit, Mr. Profit's dam, won the £50 Barham Challenge Cup at the London Dairy Show in 1897.

Mr. Cherry (6211) a descendant of that wonderful milker Young Cherry, was used in 1909-1910, as was Hermit (C.H.B. 102,494) whose dam Dorothy won three Champion Cups at the London Dairy Show in 1908. Bulls now in service are Mr. Profit, Mr. Cherry, Hermit, Burton Cup (6665) whose dam Nancy V. won both the Lord Mayor's and the £50 Challenge Cups at the London Dairy Show in 1910, and Kirmington Forester 13th (6952) the Royal and Lincolnshire winner, who was bred by Mr. George Marris.

Probably no herd has produced such successful sires in recent years, both in connection with the show-yard and the sale ring as has Mr. G. E. Sandars, of Scampton, for on ten occasions during the past 20 years they have secured the highest individual price at the Lincolnshire Red Shorthorn Association's Sales at Lincoln April Fair, and on thirteen occasions they have secured the highest average. Both in 1907 and 1908, too, Mr. Sandars won premier honours in the older bull class at the show held before the sale and in 1909, besides being second in the older class and reserve for the championship, he won in the younger class, taking both first and second prizes in the younger class in 1910.

The Lincoln Red Champion at the Royal Show at Lincoln, in 1907, and at the Lincolnshire Show at Sleaford, in 1908, was Scampton Exile (4092) who was bred by Mr. Sandars, and was sold to Mr. B. Rowland, at the Association's Sale at Lincoln, in 1904. This grand bull was by Keddington Ruby (1243), whose sons have made more money for Mr. Sandars than those of any other sire at the Lincoln sales ; and he is from a King Hal (156) dam, his grand-dam being by Hallington (135), and great-grand-dam by Cawkwell (67). Besides being the Lincoln Red Champion, he was reserved for the championship of the Lincolnshire Show in 1908, standing second only to the Royal Champion, Sir R. P. Cooper's invincible roan bull Chiddingstone Malcolm, to whom it was no disgrace to take second place. In 1909, Exile was reserved for the championship of the Lincoln Reds, but he went to the top again in 1910, and was reserve to Mr. F. Miller's roan Shorthorn, Good Friday, for the premier honours of the Show. Mr. Sandars' herd was founded at Ludford in 1885, and was thence transferred to Fillingham Manor, and so eventually in 1900, to Scampton, where it has since remained. It had a good foundation, for the principal portion of it came from Mr. S. Robson's herd in Derbyshire, where it had been taken from Cadeby (where the late Mr. William Torr, of Aylesby, pronounced it to be the best large herd he knew) in Lincolnshire, and was chiefly descended from Mr. F. Traves' herd. Other additions to Mr. Sandars' herd were made by purchases from leading breeders. Many of the cows when Mr. Sandars bought them from Mr. S. Robson, were in-calf to Cawkwell (67) a bull bred by Mr. C. B. Robson, Cawkwell, and their new owner afterwards used Ludford (172) a son of Cawkwell, who was bought by Mr. E. H. Cartwright, for use in his herd at Keddington. Some of the cows Mr. Sandars bought of Mr. N. Cartwright, Haugham, were by Mr. T. Willis' famous Windsor Benedict (C.H.B. 40,933) ; and other additions were made from the herd of Mr. Mason, of Rigsby, whose cattle were descended from the old Keddington herd. Bulls from Mr. W. Chatterton's herd at Hallington were mostly used at first, and Hallington (135) proved a most successful sire. Later on Mr. Sandars used Calceby (51) also a Hallington-bred one ; and he was followed by King Hal (156), who was by Eclipse (111) from the late Mr. W. Chatterton's famous cow Riches. Then came Great Tom of Lincoln (392) in 1896, from the Keddington herd, and a wonderful bred one, for he was by Admiral (3) from the celebrated cow Nonpareil, by Hyllus (149) ; and he was in service till 1900, when Reubens (2310) bred by Mr. Reuben Roberts, and Sotby Red 9th (516) from Stenigot Rose, joined him. Digby Conqueror (2182) from Mr. Roberts' herd also served a few cows in 1901 ; and it was in this year that Keddington Ruby (1243) made his entrance into the herd, an event which was to prove of such moment to Lincoln Red breeding. This wonderful sire was in service at Scampton right up to 1907 ; and a good bull was also used in 1906 in Sausthorpe Red 10th (4550), of Colonel Swan's breeding, he afterwards going out to South Africa where he beat all breeds at Bloemfontein for the championship of the leading show in that part of the world. Mr. Sandars in 1907, also bought for 200gs. at the Alford sales, the Royal winner, Mr. J. Langham's Brandon Grenadier (4274), and in 1907 this bull not only won in his class for Mr. Sandars at the Royal Show at Lincoln, but was reserve for the championship to Mr. B. Rowland's Scampton Exile, a bull of his own breeding. Brandon Grenadier is by Brandon Lord Chief Justice (2425), a most impressive sire ; and he is out of Brandon Nonpareil, who was by Chancellor (332) from Keddington Wainfleet (of Lord Yarborough's breeding), by Bigby (319), and has been the stud bull at Scampton till 1910.

Messrs. F. & C. E. Scorer's herd at Nettleham and Bracebridge is rapidly coming to the front in Milking Trials and Butter Tests. Mr. F. Scorer commenced to get it together in 1873, chiefly from purchases from Mr. W. Scorer, Burwell, whose herd was established in 1850. During recent years, following the excellent example of Mr. John Evens, Mr. F. Scorer has converted his cattle into a Lincoln Red dairy herd of the best type, and his success in public competitions is proving the wisdom of his action, while the proximity to Lincoln emphasizes its profitable side. Mr. Scorer and his son make high prices of their bulls, and those they purchase for use in their herd must come of the best milking families. Bonby Kinsman (4251) of Messrs. T. & W. Dickinson's breeding ; Welbourne Red Baron (3693), from Mr. J. C. Mountain's herd, and the home-bred Bracebridge Boothby 2nd (4264), Bracebridge Sudbrook No. 30 (4869) and Bracebridge Walker (4710) have all been in service at Bracebridge and Nettleham, also Kirkby Imperial (4896), Scampton Herald (4968), Laddie (6186) a bull bred by Mr. S. Crawley.

At the London Dairy Show in 1907, Mr. Scorer's Bracebridge No. 3 B. was reserve for the Lord Mayor's Cup, and, as then, was second in the Dairy Trials at the same Show in 1910. In 301 days she gave an average of 4'47 gallons a day. This cow was by Weston IXL. 2nd (2388) a bull bred by Mr. W. J. Atkinson, her dam Sudbrook No. 63, being by Withern Marshman (574) of Messrs. W. T. Wells & Sons' breeding.

The herd belonging to Mr. George Marris, Kirmington House, Brocklesby, was established at Holton-le-Moor in 1885, and was obtained from the trustees of the late Mr. T. J. Dixon (who bred largely from the blood of Booth, of Warlaby), which herd had been in existence some 60 years previous to Mr. Marris taking it over. Mr. Marris, who has been a successful exhibitor, and whose bulls always command competition at the Lincoln sales, first used Hallington blood considerably, while dipping occasionally into C.H.B. strains, and other choice Lincoln Red families ; and latterly Red Cap 2nd (3546) bred by Mr. W. Chatterton, and Scampton Forester (4557), of Mr. G. E. Sandars' breeding have been used with great success. Mr. Marris strengthened his herd by some high-price purchases at the Keddington sale in 1906, and his show-yard successes both last year and this point to further honours in the future. At the Association's Show and Sale at Lincoln in 1910, he carried off the championship with Kirmington Forester 13th (6952), a handsome bull by Scampton Forester (4557), out of Keddington Pearl, by Bigby (319), who was afterwards sold to Mr. John Evens to go into the famous Burton herd, at 82gs. For his new owner he took premier honours at the Royal and Lincolnshire Shows in 1910.

It is impossible to make anything but a very brief reference to other leading herds in so short a History, though many of them have played an important part in the Lincoln Red world. First, then, in alphabetical order, let me take the two herds belonging to the brothers Messrs. E. & H. Abraham, at Otby and Walesby, near Market Rasen, herds which they inherited from their father, the late Mr. John Abraham, of Otby, in whose hands it had been for 36 years at the time of the formation of the Herd Book, and which had been in existence since 1835. The best blood obtainable has always been used in these twin herds, and the success of the Otby and Walesby cattle in show and sale ring have been noted in previous pages.

Mr. Ekins Ashley, of Deepden Farm, Godmanchester, Hunts., who has established a herd of Lincoln Red dairy cattle, is a new-comer in the ranks ; but he is breeding on right lines, and has laid his foundation from some of the best herds, such as Messrs. S. E. Dean & Sons', Mr. A. Smith's, Mr. Everett King's, Mr. T. Bett's, and Mr. E. H. Cartwright's.

Mr. W. J. Atkinson, Weston St. Mary, Spalding, has always been a successful exhibitor at the Royal and Lincoln-shire Shows, and his private auction sales have invariably been most successful. His herd has been in existence some 20 years, and has been most carefully bred, with more frequent crosses of C.H.B. blood than in many herds. His cattle are of beautiful quality, and his bull Weston Monarch 4th (4187) won at the Royal in 1908 and was second in his class and reserve for the breed Championship at the County Show to Mr. B Rowland's Scampton Exile (4092).

A herd that is steadily coming to the front is that belonging to Messrs. T. Atkinson & Son, North Kelsey, and which has been in existence upwards of 50 years. High-priced bulls from the leading herds have always been used, and in recent years there has been a judicious use of Scampton and Bonby blood from Mr. G. E. Sandars' and Messrs. T. & W. Dickinson's herds.

Mr. C. F. Bett's herd at Benniworth was obtained by purchases at the dispersal of his father's herd in October, 1906. The late Mr. Tom Bett was a famous breeder, and his cattle has done much to make Lincoln Red history, as readers of my articles must have noticed. The cattle purchased by Mr. J. G. Williams, Pendley Manor, Tring, at the dispersal sale in 1906, did wonders for that gentleman's herd. The Benniworth herd descended to Mr. Tom Bett from his father in 1873, it having been in the possession of the latter since 1826. None but the very best blood has been good enough for Benniworth, and the selections from the Stenigot, Keddington, Hallington and Somercotes (Freshney) herds (including such as Red Chief (2611) bought at Messrs. Chatterton's sale in 1901 for 110gs.), have always been of a high character. Somercotes Bonus (4577), Stenigot Ruby Chief (3621), and Keddington Comet (3443) have all been in service at Benniworth during recent years.

A herd fast making a name for itself is that belonging to Mr. A. P. Brandt, Bletchingley Castle, Surrey, and many show yard honours have fallen to it in recent years. Founded by purchases from Stenigot at the 1907 sale, and added to by Moreton Moorel and Kirmington blood, it is being bred on right lines and must soon rank among the leading herds. Such sires as King Louis (5457) bred by Mr. S. Crawley, Oundle, and the champion bull Crimson King (5258) have been used in the herd, as well as the home-bred Bletchingley Apollo (5831), who introduces the Cropwell blood.

Messrs. S. & J. H. Briggs, Saleby, Alford, have a small but good herd; but the herds of Mr. Joseph Brocklebank, Carlton-le-Moorland, Newark, and of Messrs. C. & R. Brooks, Boothby Grange, Burgh, are bigger, all three being of considerable standing.

Mr. G. J. Brown, of Tothby House, Alford, has a herd of well over 100 females, and it is quite one of the leading strains in the county. It is well descended, and has been judiciously bred since it came into its owner's possession in 1891. Good prices have been made of the bulls, and the herd has also done well in the show ring.

Messrs. J. & G. W. Brown, Hagnaby House, Alford, are also well known as breeders of Lincoln Reds, and at the Alford and Boston sales the cattle they send all are in ready demand.

Mr. P. F. Brown, Digby, Lincolnshire, possesses a herd which was got together in 1858 by Mr. Pereira Brown, at Glentworth Hall, he using bulls bred by Mr. C. R. Fieldsend, Kirmond, Mr. G. E. Sandars, and Mr. W. Chatterton. The present owner took it over in 1899, and has continued to breed on excellent lines, using bulls of high quality from the leading herds.

Messrs. J. J. & W. B. Burtt, of Welbourn and Wellingore respectively, own herds that have been handed down to them from their grandfather, and the cattle are all descended from the " Old Welbourn Reds," which had such influence in the centre and south of the county in by-gone years.

Mr. John Byron's herd at Normanby-le-Wold was established by him in 1887, when he bought the best of his father's cattle at the dispersal sale at Goltho, a herd that had been established since 1862. Mr. Byron has never hesitated to pay a good price for sires of his fancy, and his bulls have always found a ready sale.

Miss K. Carleton's dairy herd at Gilford Castle, Co. Down, founded entirely on Burton cattle, is doing wonderfully well in the show ring and at Milking Trials.

Mr. John Crawley, Church Lawford Grange, Rugby and Mr. S. Crawley, Hemington, Oundle, are quite in the front rank of Lincoln Red breeders. Their joint herd was entered in the first volume of the Herd Book, and they have been most successful exhibitors. while their animals, whether sold at the public or their own private sales, always made good prices. One of their bulls, Bumper 2nd (1793) realized 105gs. to Mr. W. B. Swallow, at the Lincoln sales in 1900, and Crimson King (5258) was champion Lincoln Red and reserve for the championship of the Lincolnshire Show in 1909. These breeders have gone in a good deal for C.H.B. blood, though choice strains of Lincoln Red blood has been judiciously admixed.

Messrs. S. E. & J. M. Dean, Threekingham, Folkingham, have a herd that has been established since 1891, founded on sound blood, and carefully bred ever since.

No history of the Lincoln Reds would be complete without. mention of the famous herd belonging to Messrs. T. & W. Dickinson, Worlaby, Lincoln, for the part their cattle have played has been an important one indeed. It was inherited from their father in 1883, and is made up of the best and purest Lincoln Red blood, as is testified by the ready demand for their bulls at each Lincoln April Fair. Scampton Excursionist (4089), for whom Messrs. Dickinson paid Mr. G. E. Sandars a big figure at Lincoln in 1904, has proved a very successful sire in recent years, and the influence of the blood may be traced with the successes of many of the best herds of the day.

Mr. William Dods, Donington, Spalding, has a good herd that was formed in 1902 by purchases from members of the Association, and only the best blood has been introduced since.

A herd that has rapidly come into the front rank during recent years is that belonging to Messrs. J. W. Farrow & Sons, Strubby Manor, Alford, and a great amount of success has attended their cattle both in the show-yard and the sale ring. They are a massive, heavy-fleshed class of animal, and of a good colour, and they live in a county second to none. The Messrs. Farrow invariably select their sires from the leading herds, and will pay a big price for what they want.

One of the best known herds in the county belongs to Mr. George Freir, of Deeping St. Nicholas, one, like Mr. W. J. Atkinson's, largely built up on C.H.B. blood. The home auction sales have been most successful, and the handsome cattle Mr. Freir has put belore the public have made the good prices they deserved. Clan Macdonald (C.H.B. 78,597), and Dowsby Virtuoso 28th (4363), of Messrs. Dean's breeding, have proved highly successful sires during recent years ; and Mr. Freir has found that a combination of C.H.B. and L.R.S. blood has been attended with most satisfactory results, the one giving neatness in appearance and quality, while the other furnished hardiness, good constitutions, a wealth of lean flesh, and excellent milking capabilities.

Captain E. M. Grantham's herd at West Keal, Spilsby, has long been a prominent one, for its members have carried away many prizes from the Royal and Lincolnshire Shows. Colonel Grantham had been breeding it carefully since 1864, and it came into the possession of his son, the present owner, in 1896, who has turned to the best known herds for his sires.

The herd belonging to Lord Heneage, Hainton Hall, is another prominent one, and his blood is considerably sought after. His Lordship will only have the best strains imported into his herd, and latterly he has shown a partiality for the late Mr. T. B. Freshney's blood. The bulls in service in 1908 and 1909, however, were Otby Earl (5556) and Scampton Judas (6325) for whom Mr. T. Wallis paid Mr. G. E. Sandars 90gs. at the Association's sales in 1909.

Mr. Percy Hensman, Fulletby Grange, Horncastle, has come into prominence at the Royal and County Shows during the past few years with some beautiful cattle, and there is no doubt he is breeding on correct lines, and has further successes before him. ·The herd is not a very old-established one, but it has been carefully put together, and the selection of sires is a judicious one. Mr. Hensman purchased Scampton Exile (4092) from Mr. B. Rowland, and won the championship of the Lincoln Reds in 1907 and 1908, being also reserve in the latter year to Sir R. P. Cooper's Chiddingstone Malcolm for the highest honours of the show, while in 1910 he again took the breed championship, while being reserve to Mr. F. Miller's Good Friday for the championship of the yard.

The herd belonging to Mr. Everett King, of Northboro' Market Deeping and Cotterstock, Oundle, was established in 1899 by purchases from the herds belonging to Mr. Walter Martin, Messrs. Dean, Messrs. Crawley and others, and the home sales have been productive of good prices, while the success of Northboro' Cromwell 4th (2587) at the Royal and other shows stamped him as a bull of especial merit. Mr.

King is another advocate for a frequent infusion of C.H.B. blood into the herd.

The Exors. of the late Mr. George Laughton have a very good little herd at Belchford, Horncastle, well bred in every way, and fast taking a prominent place in the Lincoln Red world.

Mr. J. Mason's herd at Calceby Manor, Alford, is a very old one, as it was established by the present owner's father in 1854. It was built up largely on Hallington, Stenigot and Keddington blood, and has always been of high repute in the county, as the frequent mention of the prefix Calceby in the history of the breed goes to show. Of late years Mr. Mason has turned to Scampton strains with considerable success.

Mr. J. W. Measures has a capital herd at Dunsby, Bourne, which he took over from his father in 1881, who had been carefully breeding for many years. Mr. Measures picks his sires carefully, and likes an occasional cross with a C.H.B. bull of the right stamp and colour. In 1909, at the Association's Show and Sale at Lincoln, he carried off the championship with Dunsby Red 3rd (6017), by Sausthorpe Red 12th (4552), Mr. F. B. Wilkinson, Edwinstowe, later on giving 165gs. for him.

Col. J. W. A. Morgan, South Leverton Priory, Notts., recently founded a really good little herd on Stenigot, Deeping and other strains. A personal friend of the author, it was with great regret that he learned of Col. Morgan's sudden death in December, 1910.

Mr. J. C. Mountain's herd at Welbourn, Lincoln, is in high repute, being carefully bred since 1875, with a good strain of blood from the herd of Mr. Oliver, of Eresby, added in 1879. Mr. Mountain has always had a ready demand for his cattle.

Mr. J. B. Nelson's old-established herd at Bigby was obtained from his father in 1882, Mr. John Nelson having owned it for upwards of 25 years.

Lord A. M. A. Percy has got together a capital herd at Guy's Cliffe, Warwick, which has already made a debut in the show ring.

The herd at Healing Manor, belonging to Captain the Hon. G. B. Portman, was purchased from Mr. Maunsell Richardson in 1903, who had established it in 1889, and got together a fine herd of cattle before he left the county. Captain Portman has been using Ruby Scottish Rose (C.H.B. 96,734), and Stenigot Cornish Knight (C.H.B. 104,059) lately, and at the Association's sales at Lincoln in 1910, he paid 112gs., the highest price of the day, for Earl Fitzwilliam's Wentworth Earl (7248), a handsome son of Cropwell Dainty (4341) and Cropwell Pride IV., by Cropwell Red Earl (2851).

Capt. C. L. Prior, Grimblethorpe Hall, Lincoln, is also rapidly forming a good herd, which was started with part of Mr. J. St. V. Fox's cattle at Girsby, with additions from Mr. R. Chatterton's, Mr. George Marris', and Mr. W. A. Ewbank's (Covenham) herds.

Mr. Reuben Roberts' herd has long been one of the most noted among the Lincoln Reds, for its owner had been carefully building it up for some 28 years before the Shorthorns of the county became a registered breed, and he has always gone to the best blood for his out-crosses, sparing no expense. It was founded on cattle bought from Mr. Mason, of Orgarth Hill, Tathwell, and Mr. S. Welfitt, of Tathwell Hall, both of whom had been breeding Red Shorthorns for many years; and the records of the sale-ring will show in what repute Mr. Roberts' cattle has always been held. Beginning with Stenigot bulls at the time of the formation of the Herd Book, he afterwards introduced some Calceby, Benniworth, and Scampton blood, and later on brought in some choice strains from the Keddington herd; and the bulls in service during recent years were Bonny Boy (3758), bred by Mr. J. W. Measures; Scampton Excelsior (4085) a very successful sire of Mr. G. E. Sandars' breeding; and Strubby Champion (5005) bought from Messrs. J. W. Farrow & Sons; while Bracebridge Baron II. (5173), of Mr. F. Scorer's breeding, entered the herd in 1908.

A very old-established herd is that belonging to Messrs. J. W. Robinson & Sons, Anderby, which has been handed down from father to son since 1800, and very few heifers have been added to the herd. Stenigot and Hallington blood has been chiefly used by Mr. J. N. Robinson, who obtained the herd from his father in 1850, and only sires from herds of repute have been used ever since, the Keddington and Calceby strains being most in favour. There is ever a ready sale for the Anderby bulls at Alford and elsewhere.

The Croft herd goes back to 1817, when Mr. Henry Searby took the Crown Farm, at Croft, near Wainfleet. Mr. John Searby, the present owner, took it over from his father in 1885, and it has always been one of the leading herds. Various additions have been made from prominent herds, and the bulls for service have always been carefully selected. The famous Keddington Ruby (1243) was at one time in service at Croft, and the blood of Imperial Favourite (C.H.B. 86,233) has recently been introduced through one of his sons, bred by Messrs. S. E. Dean & Sons. Since 1907 the blood of Croft Aubourne (4325), Croft Son of Violet (4791), Saltfleet Exchange (5602), Scampton Justinian (6335) (which after winning for Mr. G. E. Sandars in the younger bull class at the Association's Show and Sale in 1909, cost Mr. Searby 100gs. when he came under the hammer), and Threekingham Favourite (5034) has been used in the herd.

One of the best known herds on the East Coast of Lincolnshire is that belonging to Mr. B. Simons, Willoughby Grange, Alford, which has turned out some rare cattle, and played an important part in Lincoln Red breeding. Bulls from the best known herds have invariably been used, and a capital strain of C.H.B. blood was introduced through Lord Augustine (81,472).

A herd of note is that belonging to Mr. L. W. Stephenson, South Thoresby, Alford, which was established by the present owner in 1866, mostly of Mr. Coulam's strains. Since then Mr. Stephenson has judiciously introduced Keddington, Hallington and Stenigot blood ; and at the Association's sale at Lincoln in 1905, he paid 100gs. for Mr. G. E. Sandars' Scampton Fortress, who has proved an impressive sire.

Mr. W. B. Swallow, Wootton Lawn, Ulceby, and Horkstow, Barton-on-Humber, has proved a most enterprising and go-ahead breeder. An old-established herd, for it was founded by his uncle the late Mr. J. B. Swallow in 1878, who obtained it from Mr. Percival Richardson, of Horkstow, in whose family the herd had been for several generations, the present owner has spared no expense in obtaining the best possible sires. For instance, in 1900, at the Association's April Bull Sale, he paid 105gs. for Messrs. S. & J. W. T. Crawley's Bumper 2nd (1793), a grand bull by Baron Ormsby 3rd (26) ; and at the 12th Sale of the Association in 1907, he gave 140gs. for the first prize-winning bull Scampton Hermes (4972), bred by Mr. G. E. Sandars, a splendid specimen of a Lincoln Red, by Keddington Ruby (1243) from a Digby Conqueror (2182) dam. Mr. Swallow has also introduced the late Mr. George Walker's Bigby blood, as well as Hallington and Cropwell strains with most satisfactory results.

A carefully-bred herd, and a successful one in the showyard is that bred by Mr. John Todd, Kirkby Green, which has been in the owner's possession since 1887. The frequent reference to the prefix " Kirkby " to the heroes and heroines of the show and sale ring speaks for the standing of the herd.

Mr. John Tomlinson's cattle from Birthorpe Manor are also coming to the front ; while Mr. E. J. Turton's herd at Horkstow, which has been in existence there since 1878, and was for 40 years before that carefully bred by Mr. John Turner, at Ulceby, is one of the best known in North Lincolnshire.

The herd founded by Mr. F. B. Wilkinson, Edwinstowe, Newark, in 1906, has proved a most successful one in the show-yard, and as Mr. Wilkinson is a most determined and enterprising bidder for the very best in the sale ring, it will go further still towards the front.   Mr. Wilkinson purchased the champion of the Association's show and sale at Lincoln, in 1909, Dunsby Red 3rd (6017), for which he paid 165gs., and he has since used the good bull Hallington Neptune (3904) and Saltfleet Friar (5603) both well known in the show-yard.

One of the most successful herds in the show-ring during recent years is that belonging to Mr. J. G. Williams, Pendley Manor, Tring, which was founded in 1899 by purchase sfrom the famous herd belonging to Mr. John Evens, at Burton. Additions were afterwards made from the herds of Messrs. Crawley, of Hemington and Oundle, and Messrs. Chatterton, of Stenigot, while some of the best cows and heifers were bought at the sales of Mr. E. H. Cartwright, Keddington, and Mr. T. Bett, Benniworth ; and as the Scampton blood was introduced through Blood Stain, who was used in 1901, 1902 and 1903, it will be seen of what rare families the herd is composed.   Mr. Williams' wonderful successes at the Royal, Lincolnshire, Peterborough and Tring Shows, by means of his judicious purchases and their offspring, will have been noticed in my previous articles.

Among the good herds that have been dispersed, but have played no small part in the building up of the Lincoln Reds, the most prominent of those of which no mention has already been made was that belonging to Messrs. S. E. Dean & Sons, Dowsby Hall, Bourne and Heath House, Nocton, Lincoln, who have long been in the forefront of the breeders and exhibitors of Lincolnshire Red Shorthorns, and their enterprise has had much to do with the improvement in the appearance of the cattle in recent years.   Their herd was founded in 1866 by Mr. S. E. Dean, and came into the possession of Messrs. J. H. & A. W. Dean in 1881, and, largely built up on C.H.B. blood, was most carefully maintained ever since.   These gentlemen attended Mr. W. S. Marr's dispersal sale and purchased the bull calf Imperial Favourite (C.H.B. 86,233) for

700gs. and afterwards sold him at Buenos Ayres for the equivalent of 1,000gs. He was most extensively used in the Dowsby and other herds ; and so was his son Dowsby Imperialist (C.H.B. 91,507), and later on Baron Broadhooks (C.H.B. 90,785).

The Brandon Grange herd, belonging to Mr. J. Langham, was once one of the most successful in the show ring. Mr. Langham got it together in 1888, and he was successful in breeding several notable animals such as Brandon Lord Chancellor (2121), Brandon Lord Chief Justice (2425), Brandon Grenadier (4274) (for whom Mr. G. E. Sandars paid 200gs. at Alford Fair), Brandon Nonpareil (3252), and Brandon Christmas Gift (3256). A perusal of the chapter dealing with the breed in the show ring will demonstrate the prominent position of the herd in the Lincoln Red world.

Other noted herds that have been dispersed were those of the Earl of Yarborough, at Brocklesby Park ; Sir Robert Wilmot's, at Binfield Grove, Bracknel ; Mr. J. Maunsell Richardson's, at Healing Manor ; Mr. C. H. Stafford's, at Fledborough ; and the late Mr. A. Smith's, of Surfleet, though his sons are still breeding Lincoln Reds on the sound lines he laid down.

Printed by J. W. Ruddock & Sons, Lincoln.

KIRMINGTON FORESTER 13TH (6952),

The property of Mr. JOHN EVENS, Burton, Lincoln,

Bred by Mr. G. MARRIS, Kirmington, Brocklesby,

First and Champion at L.R.S.A. Show at Lincoln ; First at the Royal and Peterborough Shows ; First and Reserve Champion at the Lincolnshire Show, 1910.

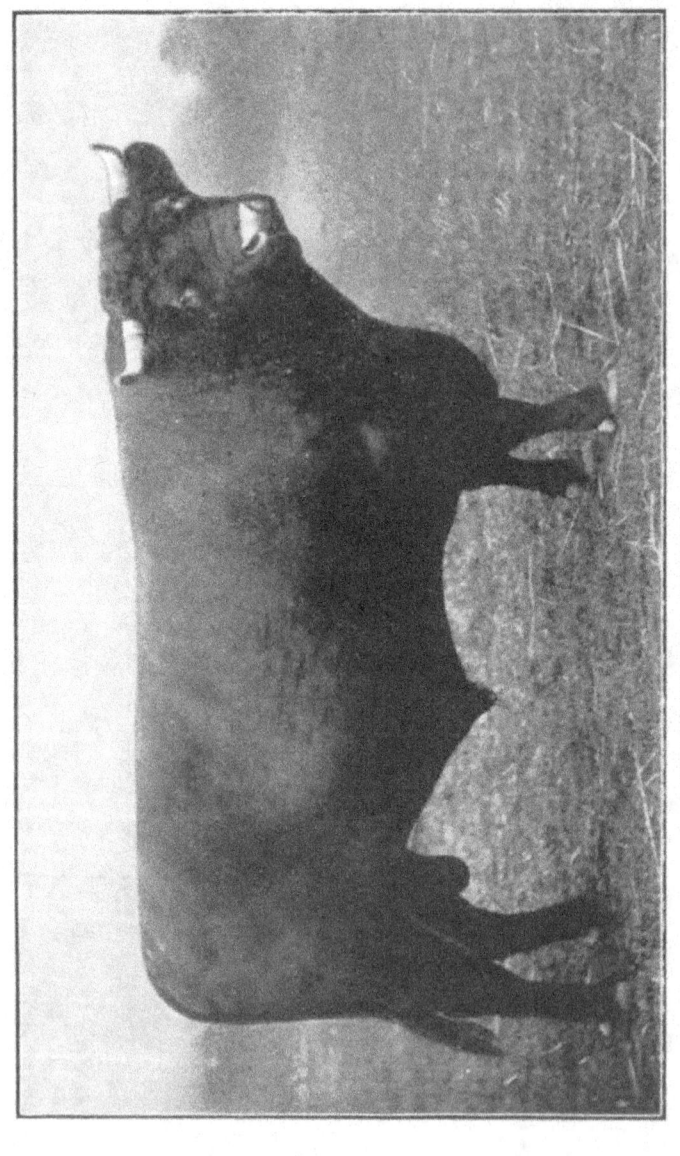

SCAMPTON EXILE (4092),

The property of Mr. P. J. Hensman, Fulletby Manor, Horncastle,

Bred by Mr. G. E. Sandars, Scampton, Lincoln,

Champion Lincolnshire Red Shorthorn Bull in 1907, 1908, and 1910.

BRACEBRIDGE No. 3B.

Bred by and the property of Mr. F. SCORER, Bracebridge Heath, Lincoln,
Second Prize and Reserve for Lord Mayor's Cup at the London Dairy Show in 1907,
and Second in 1910.

www.ingramcontent.com/pod-product-compliance
Lightning Source LLC
Chambersburg PA
CBHW082012230526
45468CB00022B/1978